Centrifugal Pumps for Sodium Cooled Reactors

This comprehensive introduction to centrifugal pumps used in sodium-cooled fast reactors discusses the special attributes of centrifugal pumps, design features, manufacturing requirements, instrumentation, and operating experience. It covers the characteristics of mechanical pumps, used as the main coolant pumps in fast reactors.

Key Features

- Covers description of pumps in various reactors highlighting the special features of the pumps and providing an overview of futuristic design concepts.
- Discusses the aspects related to the design, manufacture, testing, instrumentation, and operating experience of centrifugal sodium pumps.
- Highlights the challenges in centrifugal sodium pump testing.
- Presents topics such as cavitation testing for critical applications and thermodynamic effect on pump cavitation.
- Real-life case studies are included for better understanding.

This book gives a detailed overview of the design, manufacture, testing, and operating experience of the main coolant pumps used in sodium-cooled nuclear reactors. It further discusses the special type of pumps used in fast reactor power plants to circulate liquid sodium through the core. The text examines the challenges in centrifugal sodium pump testing and types of test facilities around the world. Real-life examples are used to highlight important aspects. It is primarily written for senior undergraduate, graduate students, and academic researchers in the fields such as mechanical engineering, nuclear engineering, and chemical engineering.

Centrifugal Pumps for Sodium Cooled Reactors

R. D. Kale
B. K. Sreedhar

CRC Press
Taylor & Francis Group
Boca Raton London New York

CRC Press is an imprint of the
Taylor & Francis Group, an **informa** business

Front cover image: Indira Gandhi Centre for Atomic Research

First edition published 2024
by CRC Press
2385 NW Executive Center Dr, Suite 320, Boca Raton, FL, 33431

and by CRC Press
4 Park Square, Milton Park, Abingdon, Oxon, OX14 4RN

CRC Press is an imprint of Taylor & Francis Group, LLC

ISBN: 978-1-032-46053-6 (hbk)
ISBN: 978-1-032-60735-1 (pbk)
ISBN: 978-1-003-46035-0 (ebk)

DOI: 10.1201/9781003460350

Typeset in Sabon
by SPi Technologies India Pvt Ltd (Straive)

To the memory of my late wife Mrs. Lata Kale, and my sons, Udayan and Kedar.

— R. D. Kale

To the loving memory of my late parents Kongot Balagopalan and Kondapurath Devayani, whose unwavering love and support continue to inspire me.

— B. K. Sreedhar

Contents

Foreword

This book on *Centrifugal Pumps for Sodium-Cooled Reactors* by Mr. R. D. Kale and Dr. B. K. Sreedhar is an important contribution towards making an authentic, experience-based resource available to practitioners, R&D personnel, academics, and students interested in the field of centrifugal pumps for pumping sodium and similar fluids. This is a highly specialised area where there is a paucity of literature, and this book would help bridge the void in a significant way.

India, moving ahead with indigenous development of advanced technologies in critical areas, is a matter of high importance from the perspective of avoidance of vulnerabilities at the national level, maintaining a strategic edge, and of course, domestic value addition. Our Atomic Energy, Space, and Defence programmes have been engaged in mastering several such critical technologies. A number of agencies outside these programmes, including industries and academia, are deeply connected and involved in such efforts. It is necessary that there is growth and continuity in terms of HRD activities in these areas across different organisational domains. Books like this serve as important resource materials for such purposes.

In particular, the development of dynamic equipment such as a pump is full of significant challenges. Apart from the development of materials, design of various parts as well as the system as a whole, and their manufacture into required shapes for various components conforming to required specifications, such efforts also involve the design of several kinematic pairs working together. A successful development thus necessitates deeper insights into issues related to disciplines like hydrodynamics, structural dynamics, tribology, manufacturing, and materials. Further, the design and development of pumps meant for nuclear reactors, that too for pumping very reactive fluids like sodium, need extreme care and qualification in terms of their functional and safety performance. The availability of a book written by authors involved in developing this kind of equipment in the country would fulfil a much felt need of younger professionals who might want to get into this area. The value of this book is even greater considering that sodium-cooled fast reactors are expected to be set up in large numbers as a part of the second stage of India's nuclear power programme, and we expect industry and academia also to be a part of this process in addition to work within the Department of Atomic Energy.

I wish to compliment the authors for their efforts in bringing out such a valuable book and feel confident that it would be an important addition to the limited literature on the subject.

Anil Kakodkar
Former Chairman Atomic Energy Commission, India

Preface

More than two decades ago, the first author, while in service at the Indira Gandhi Centre for Atomic Research (IGCAR), Kalpakkam, had compiled a document with his colleagues titled "History Document on Design and Development of Primary Sodium Pump".

This document detailed the efforts invested in the development of the primary sodium pump for the Prototype Fast Breeder Reactor (PFBR) project. At that time, he had also prepared another document called "Design Manual of Primary Sodium Pump" for the benefit of the engineers who would be entrusted with the construction and commissioning of this component. The roots of this book lie somewhere in these works.

Although there is a library of books on the topic of centrifugal pumps, it is surprising that there is no book devoted to mechanical or centrifugal sodium pumps. We have come across only a single book on the topic of nuclear reactor coolant pumps in the Russian language. This book includes a brief discussion on sodium coolant reactor pumps. Unfortunately, no English translation of the book has been published, and the original itself is now hard to come by.

The present book, *Centrifugal Pumps for Sodium-Cooled Reactors*, consists of ten chapters. Unlike other published books on centrifugal pumps, which focus on the calculation of design parameters, discuss pump characteristics, selection of bearings, seals, drives, general pump troubleshooting, and so on, we focus instead on issues that a sodium pump designer is confronted with when he embarks on the journey from scratch. For instance, the chapter on design discusses the choices available to the designer on the location of the pump, the hydraulics, the mechanical aspects, etc. Abundant examples of design choices from pumps of various reactors are provided to reinforce a technical point.

The book, in its ten chapters, provides a comprehensive review of the pump designs used in fast reactors around the world, the options available to the designer, the challenges in manufacture and in testing, and the operating experience of these pumps over a span of more than half a century.

Each chapter is provided with a reference list for the benefit of readers who wish to study a particular topic in further detail. Another feature of the book is the presentation of details of pump components, such as special bearings, shaft sealing, and pump supports. These are supplemented with neat sketches to aid in a better understanding of their complexities. We also include a chapter that discusses, in brief, the main coolant pumps of water-cooled reactors. The purpose of including a discussion on water-cooled reactor pumps is to bring together important topics about these pumps to benefit readers conversant with water-cooled reactors but with little access to information on such pumps.

The authors believe this chapter will interest engineers working in nuclear plants in India, in particular.

In the final chapter, considering the ongoing R&D efforts worldwide, we attempt to predict the features of a sodium pump of the future.

The authors have many years of experience in the field of nuclear engineering, with a particular focus on the design and testing of centrifugal pumps for fast reactors and test loops. They draw on this wealth of experience to carve out a lucid picture of the design, manufacture, testing, instrumentation, and operation of centrifugal sodium pumps over the last 60 years and collate the details in an easy-to-understand manner between cover and cover.

Knowledge in any field, especially technical, is gained not just from formal education or perceptive wit. Equally important is our understanding of the experience of others and our own. This sublime Sanskrit shloka eloquently captures the essence of the learning process:

आचार्यात् पादमादत्ते पादं शिष्यः स्वमेधया ।
सब्रह्मचारिभ्यः पादं पादं कालक्रमेण च ॥
AchAryAt pAdamAdatte, pAdam shiShyaH swamedhayA |
sa-brahmachAribhyaH pAdam, pAdam kAlakrameNa cha ||

In translation:
From the teacher one-fourth is learnt, one-fourth using one's intelligence, one-fourth is learnt from colleagues, and one-fourth only with time.

The authors are indebted to Dr. Anil Kakodkar, former Chairman, Atomic Energy Commission (AEC), and Secretary to the Department of Atomic Energy (DAE), and presently member AEC, for graciously penning the foreword to the book.

Acknowledgements

The authors acknowledge the contributions of their former colleagues, Messrs A.S.L Kameswara Rao, S. Baskar, K. Balachander, Chander Raju, late S. Asok Kumar, K.V Sreedharan, P. Ramalingam, and R. Prabhakar, towards the design and development efforts for the main coolant pumps of PFBR.

The authors also appreciate the pioneering contributions of Kirloskar Brothers Limited, Pune (KBL), India, and, in particular, Mr. S.G. Joshi, Hydraulics expert and formerly Vice-President, KBL, Pune, in the development and manufacturing efforts of the main coolant pumps of PFBR.

The authors are grateful to the Director, Indira Gandhi Centre for Atomic Research (IGCAR), Kalpakkam, India, the management of IGCAR, and the Department of Atomic Energy, Mumbai, India, for permission to publish this book.

Many people have helped in the writing of this book. The authors place on record their sincere thanks to the following colleagues:

Mr. A. Kolanjiappan for his dedicated efforts in producing the sketches for the book.
Ms. S. Saravanapriya for typing the initial drafts of some of the book's chapters.
Mr. K.V Sreedharan for sharing technical information and clarifications on technical issues.
Mr. S. Chandramouli for clarifications on sodium loop operation.
Dr. G. Vaidyanathan for his guidance on issues related to publishing and obtaining copyright permissions.

While utmost efforts are made to weed out errors, the authors remain wary of the venerable Edward A. Murphy Jr.'s eponymous law. Readers are welcome to email the authors about any inconsistencies/mistakes in the book, and the corrections will be done at the earliest opportunity.

R. D. Kale, *Pune*
Dr. B. K. Sreedhar, *Kalpakkam*

Authors

Mr. R. D. Kale graduated in Mechanical Engineering from the Indian Institute of Technology (IIT), Kanpur, in 1966. He completed a one-year orientation course in Nuclear Science and Engineering as part of the 10th batch of the Training school at the Bhabha Atomic Research Centre and started his career in BARC. He was deputed to CEN de Cadarache, France, where he worked along with French specialists as part of the Indian design team for the design Fast Breeder Test Reactor under a collaborative project with the French atomic energy establishment from May 1969 to June 1970. He moved to the Indira Gandhi Centre for Atomic Research (IGCAR), Kalpakkam, in 1971 and worked for a few years initially on the detailed design of some reactor components and preparation of their technical specifications.

He spent the next 28 years establishing water and sodium test facilities to test fast reactor components. These included a large sodium test facility in a 43 m high building for testing full-scale reactor components in sodium and later a dedicated Steam Generator Test Facility for testing a sector prototype of the once-through sodium/water steam generator of the 500 MWe sodium-cooled power plant. He spearheaded the indigenous development of sodium centrifugal pumps for the Prototype Fast Breeder Reactor (PFBR) that culminated in the manufacture of both the primary and secondary centrifugal pumps by the Indian industry. He has several publications in reputed journals, seminars, and international conferences. He has published many articles in newspapers on nuclear power and safety. He retired in the grade of Outstanding Scientist. He was the Associate Director of the Engineering Development Group and Director of the Engineering Services Group at the time of superannuation. Post retirement he worked as Member, Project Design Safety Committee of the Atomic Energy Regulatory Board (AERB) for PFBR.

He has also authored a book in Marathi titled *Adhunic Hindu Dharma, Kaani Kasa?* translated as *Modern Hinduism, Why and How?*. In press, his recent book in Hindi is titled *Meri Yatra, 1857 kaAnkhoDekha Hal*, translated into English as "*My Journey, an eyewitness account of the 1857 freedom struggle*". The book is a translation of the Marathi title

Maza Pravas that traces a priest's journey from Penn Tahasil near Mumbai to Jhansi and other holy places in North India in the backdrop of tumultuous events in the first war of Indian independence in 1857.

He is a fellow of the prestigious Indian National Academy of Engineering (INAE), member of the Institution of Engineers, India, member Indian Institute of Chemical Engineers, and a member of the Indian Nuclear Society.

Dr. B. K. Sreedhar is in the grade of Outstanding Scientist and is the Director of the Fast Reactor Technology Group at IGCAR, Kalpakkam. He graduated with a gold medal in Mechanical Engineering from the University of Calicut in 1989. He completed the year-long orientation course in Nuclear Engineering as part of the 33rd batch of the BARC Training School and joined IGCAR in 1990. He started his career in the indigenous hydraulic development of primary and secondary coolant pumps for the PFBR. He was later involved in the in-sodium testing of full-scale mechanisms/ machines of PFBR. He is also engaged in development work towards realising oil-free pumps using active magnetic bearings and ferrofluid seals.

He has a postgraduate degree from the Indian Institute of Technology (IIT), Madras specialising in Hydroturbomachines, and a doctorate from the Homi Bhabha National Institute (HBNI), Mumbai. He has published several papers in refereed journals and national/international seminars and conferences.

He is a fellow of the prestigious Indian National Academy of Engineering (INAE), fellow of the Institution of Engineers, India, and a member of the Indian Nuclear Society.

Chapter 1

Introduction to centrifugal sodium pumps in fast reactors

1.1 INTRODUCTION

Fast neutron reactors constitute the critical second stage in India's three-stage nuclear power programme.

1. In the first stage, Pressurised Heavy Water Reactors (PHWR) fuelled by natural uranium are employed to generate power. These reactors are known as thermal reactors because the average energy level of the neutrons is equal to that of the atoms of the surrounding medium (at ordinary temperatures, the average energy of thermal neutrons is ~ 0.04 eV). In PHWR, the fissile isotope uranium-235(U235) (0.7% in natural uranium) present in natural uranium undergoes fission to produce power. The process also generates a small fraction of plutonium-239(Pu239) by the neutron absorption of uranium-238(U238) (which isotope constitutes the bulk of natural uranium).

2. Pu239 recovered from the first stage is used along with depleted uranium, mainly U238 (mixed oxide fuel), to produce power in fast neutron reactors that mark the second stage. These reactors are known as fast reactors because the energy level of the neutrons is in the range of 10–100 keV. In particular, the fission of Pu239 not only produces power but the accompanying high yield of neutrons results in the conversion of fertile U238 to fissile Pu239, thus producing more fuel than is consumed (these reactors are therefore also known as 'breeders'). This stage is critical due to the potential of generating power while producing more fissile material (Pu239) and because the technology, once perfected, can convert fertile thorium-232(Th232) (abundantly present in India) to fissile uranium-233(U233).

3. Fissile U233 from the second stage fuels reactors (both thermal and fast reactors) to generate electricity, thus utilising the large thorium-232 reserves in India. Table 1.1 [1] shows the potential power generation in the three stages. The importance of fast reactors to India's energy security is evident from the electricity potential of FBR's and Thorium breeders.

DOI: 10.1201/9781003460350-1

Table 1.1 Enhancing electricity potential with fast breeder reactors

Fuel type of reactor	Quantity tons	Electricity potential GWe-yr
Uranium – metal	61,000	
In PHWR		328
In FBR		42,200
Thorium – metal	225,000	
In breeders		150,000

1.2 FAST REACTOR LAYOUT AND ITS INFLUENCE ON PUMP DESIGN

Fast Reactors do not use a moderator to slow down the neutrons produced during fission because the neutron yield is high when Pu fissions with high-energy neutrons (up to 2.9 neutrons/fission) [2]. Furthermore, the parasitic cross-sections of materials are low at high neutron energies. On the other hand, the fission cross-section of fertile U238 is significant at high neutron energies, enabling the generation of more fissile material (by transmutation of U238 to Pu239) than is consumed. The absence of a moderator makes these reactors of high-power density[1] when compared to heavy water and light water reactors. Therefore, the coolant used in these reactors should possess good heat transfer characteristics coupled with minimal neutron energy moderation, thus limiting the choice of coolant to liquid metals. Sodium is the preferred choice among liquid metals because of its high thermal conductivity (100 times higher than water), high boiling point, moderate specific heat and low neutron energy moderation.

There are two different configurations in the layout of a fast reactor: loop-type concept and pool-type concept. In a loop-type reactor, the reactor vessel, containing the core immersed in a pool of sodium, the intermediate heat exchanger (IHX), and the main sodium pump are connected by piping of the primary circuit (Figure 1.1). The pump is located in a separate tank adjoining the reactor vessel, with the pump suction in contact with the liquid in the tank and the pump discharge joined to the system piping (sump type concept) or with the pump suction linked to the system piping and the pump discharge connected to the sump (piped concept). Since the primary pump is located outside the reactor pool, the designer has the option of locating it in either the hot leg or the cold leg. The run of piping from the reactor vessel outlet to the inlet of the IHX is the hot leg, while that connecting the IHX outlet to the reactor vessel inlet is the cold leg.

In contrast, the pool-type concept (Figure 1.2) houses the reactor core and the entire primary circuit and its components (pump and IHX) in a pool of sodium. In the pool-type reactor, since the primary pump is immersed in the sodium pool, the sizing of the pump directly affects the size of the main vessel and, therefore, the capital cost of the reactor. However, the pool type concept restricts the option of locating the primary pump to the cold pool only.

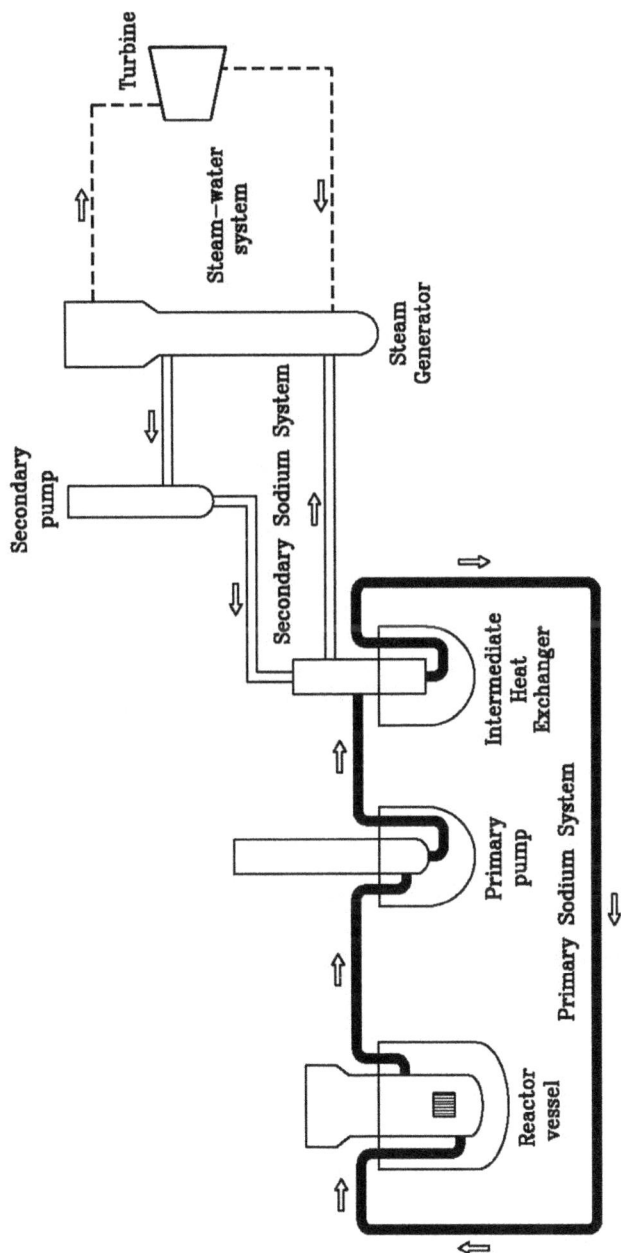

Figure 1.1 Loop-typeconfiguration.

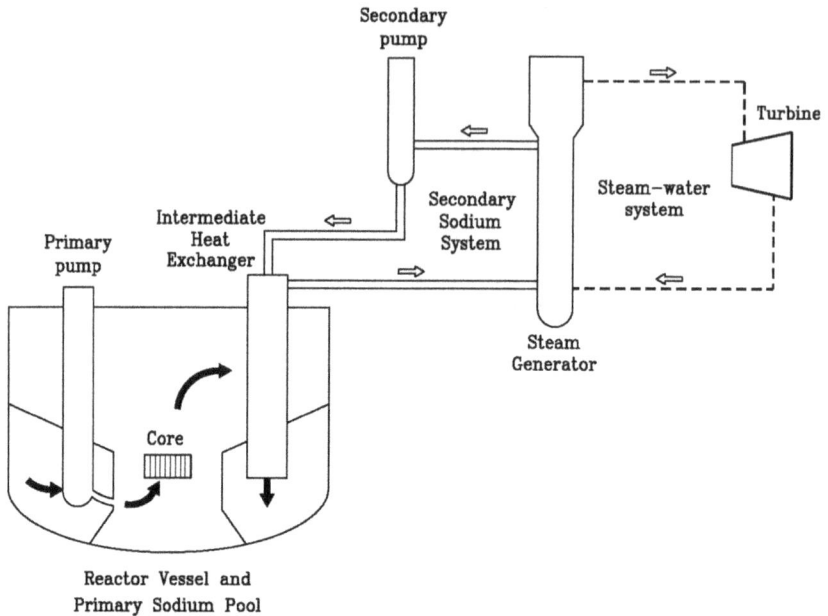

Figure 1.2 Pool-typeconfiguration.

The advantages and disadvantages of locating the pump in the cold leg are discussed in Chapter 2. Table A1.1 in Appendix 1 gives the types of fast reactors designed/installed/operated.

1.2.1 Types of pumps for liquid sodium pumping

Essentially, two types of pumps, mechanical and electromagnetic, are used to pump liquid sodium in the various reactor cooling circuits. In the mechanical pump, the pressure to circulate the coolant is produced by rotodynamic action. The rotation of the impeller, mounted on the pump shaft, imparts kinetic energy to the liquid which is transformed to pressure energy in the stationary diffuser.

Liquid sodium is pyrophoric and is to be perfectly sealed from the atmosphere. In the mechanical pump, sodium is sealed from the atmosphere using either a hermetically sealed arrangement or the shaft emerging from the pump casing is sealed at the shaft-casing interface using a seal. The canned motor pump is an example of the hermetically sealed arrangement. The rotor assembly, comprising the impeller mounted on the rotating shaft, is immersed in sodium. The rotor (armature) is shielded from the pumped liquid by a thin sheeting, while the stator is protected from the pumped liquid by a thin can of negligible electrical conductivity welded to the pump body. The rotor is supported by sleeve bearings at either end of the armature.

A fraction of the liquid from the pump discharge is diverted through the clearance between the rotor and stator, cooling them in the process; the other stream of liquid is used to lubricate the sleeve bearings. Both streams return to the impeller suction. The EBR-II primary pumps, manufactured by Byron Jackson Pump company for Atomics International, are canned motor units. In this particular design, the pump casing is integral to the pump tank, which contains sodium topped by inert gas – the pump shaft passes through a gas-tight housing consisting of a labyrinth seal. The motor is enclosed in a sealed housing, and the gas pressure in the housing is equalised with that in the liquid tank, thus avoiding the need for a mechanical seal. The electrical connections to the motor penetrating the motor housing are seal welded. The motor bearings are conventional, lubricated bearings, and the bearing in sodium is of hydrostatic type [3].

In the case of pumps with seal, direct sealing of sodium from the atmosphere is avoided by making the pump of vertical construction. A vertical arrangement permits free level of sodium in the pump barrel topped by inert (argon) cover gas. The inert cover gas acts as a buffer fluid between the free surface of sodium in the pump and the atmosphere, easing the difficulty in sealing. The sealing of the cover gas space from the atmosphere is achieved by means of mechanical seals mounted on the shaft. A repair seal for the shaft is additionally provided to isolate the pump internals, during replacement of the main seal, without disturbing the pump from the circuit. This design is the most popular and almost universally used for both primary and secondary sodium pumps. Mechanical pumps are also classified on the basis of the length of the shaft as either long shaft pumps or short shaft pumps. Long shaft pumps are provided with a bearing in the pumped liquid, whereas short shaft pumps have overhanging impeller. Most of the pumps used in sodium-cooled reactors are of the long shaft type (e.g., RAPSODIE, SuperPhénix, PFR, FBTR, PFBR). Examples of short shaft pumps are the main and auxiliary sodium pumps of the Sodium Reactor Experiment (SRE) reactor.

The electromagnetic pumps (EM) operate on the principle of interaction of current carrying conductor (in this case, the liquid metal) with a magnetic field. Compared to mechanical pumps, the main advantage of these pumps is the freedom from penetration of the pump boundary (such as is required for the shaft in a rotating equipment) and the absence of moving parts. This simplifies the design and makes the pumps almost maintenance free. EM pumps are broadly divided into conduction pumps and induction pumps. Conduction pumps are further classified into DC conduction Pumps and AC conduction pumps. Example of DC conduction pump used in a fast reactor is the primary pump of the EBR-I plant. Induction pumps are classified as AC linear and AC rotary pumps. The Flat Linear Induction Pump (FLIP) and Annular Linear Induction Pump (ALIP) belong to the category of AC linear pumps. The largest AC linear induction pump (FLIP) used in a sodium-cooled reactor is that used in the EBR-II secondary circuit.

In India, annular linear induction pumps have been successfully developed up to a capacity of 170 m³/h at a pressure of 4 bars. Use of large capacity EM pumps is limited because of their rather low efficiency,[2] possible instability from magnetohydrodynamic effects and absence of inherent inertia required for coolant flow coast down in the event of pump de-energising (trip). The economic trade-off of using EM pumps instead of mechanical pumps could be significant if improvements in these areas are realised. Table A1.1 in Appendix 1 summarises the type of pumps used in various experimental, prototype, and commercial reactors.

The main circulating pumps in a fast reactor plant are complex units with dedicated auxiliary systems, special instrumentation and many unique technical features (see Figure 1.3).

1.3 DESIGN CONSIDERATIONS FOR MAIN COOLANT PUMPS (MCP) OF FAST REACTORS

Pumps are the heart of a power plant; more so in sodium-cooled reactors what with the high-power density of the core, the harsh operating conditions, and the reactive nature of the liquid sodium coolant. The economic penalties resulting from downtime of these large power plants demand high reliability of the main coolant pumps and make the design and operation of these pumps a challenge. The design considerations of centrifugal pumps for sodium application are dictated by the unique operating environment and so these pumps have design features that are not common in conventional pumps.

(a) Material of construction: The material used for construction is to have good strength at high temperature as well as good machinability and weldability. It is to be chemically resistant to sodium, decontaminating alkali/acid solution and steam/water washing media. The material used for construction of hydraulic parts is to have good resistance to erosion at high sodium velocities and particularly to cavitation damage. As a rule, austenitic stainless steel with various kinds of thermal or thermochemical treatment is used. Hardfacing is employed in parts that require enhanced hardness. Special care is exercised in manufacture which includes raw material inspection, casting, fabrication, machining and heat treatment (e.g., of the long rotating shaft), quality control and inspection after manufacture, and lastly balancing of the rotating assembly.

(b) Compactness: The sizing of the pumps have a strong influence on the main vessel size (in the case of pool-type reactors) and the reactor capital cost. Hence the design is optimised to permit operation at the maximum speed possible. However, this aggravates the danger of cavitation during operation because the Net Positive Suction head (NPSH)[3] available is modest.

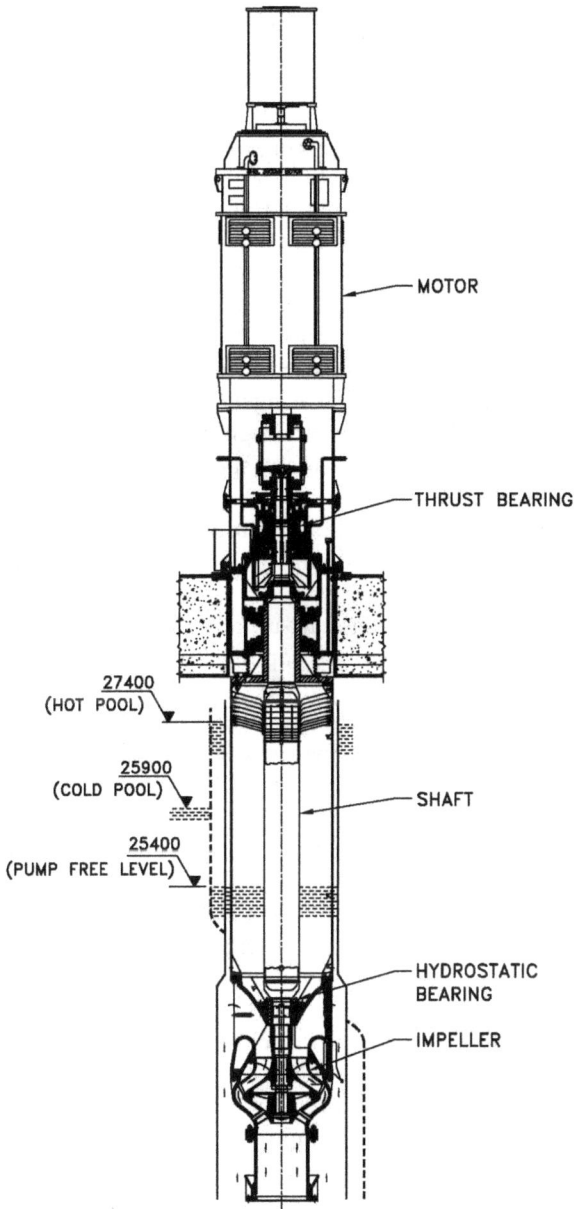

MOTOR

THRUST BEARING

27400
(HOT POOL)

25900
(COLD POOL)

25400
(PUMP FREE LEVEL)

SHAFT

HYDROSTATIC
BEARING

IMPELLER

Figure 1.3 Primary sodium pump of the Prototype Fast Breeder Reactor (PFBR).

(c) Long shaft with composite construction: The shaft is normally supported by two bearings one within sodium and the other in the cover gas space. In order to achieve maximum submergence a long shaft is used thus increasing the span between the bearings resulting in lowering of the critical speed of the rotating assembly. The shaft is therefore fabricated by welding together solid and hollow sections in order to raise the critical speed above the operating speed.

(d) Bearing under sodium: Conventional grease lubricated, antifriction bearings cannot be used to support the rotating assembly under sodium. Therefore, hydrostatic bearing which uses the high-pressure liquid feed from the pump discharge is used to support the rotating assembly. Since the load bearing capacity of the bearing is proportional to the pump speed there is some rubbing of rotating and stationary parts of the bearing during start-up/final stage of coastdown of the pump. These surfaces are therefore hardfaced to ensure adequate wear resistance. The bearing is designed to meet the following requirements:

  (i) Have minimal wear of bearing working surfaces for specified life-time taking into account the significant number of starts/stops.
  (ii) Permit pump operation at any speed within the boundaries of the working envelope.
  (iii) Allow for reverse rotation as far as possible.
  (iv) Consume a minimal quantity of pumped liquid for feeding the hydrostatic bearing.

(e) Quick acceleration during start-up and gradual coasting down: The design of the motor permits rapid acceleration during start-up so that the duration of rubbing at the under-sodium hydrostatic bearing surface is minimal. A flywheel is provided on the rotating assembly to provide adequate rotational inertia and achieve gradual coasting down of the pump flow rate in the event of motor trip under emergency conditions.

(f) Non-return valve: The primary radioactive system is provided with multiple pumps operating in parallel to guarantee redundancy and ensure adequate core cooling in the event of single pump failure/loss of power supply. Under such conditions, the flow rate delivered by the operating pumps gets partly diverted through the stopped pump bypassing the core. This bypassing of the core is prevented by using a non-return valve at the pump discharge (e.g., in the RAPSODIE primary pump), which acts as a "flow diode" permitting flow through the pump in only the forward direction.

(g) Seal: Leak tightness of the pump is paramount because liquid sodium catches fire on exposure to air. The task of sealing is simplified by making the pumps of vertical construction with the free surface of sodium topped by inert (argon) cover gas and sealing the cover gas, instead of sodium, from the atmosphere. Moreover, all joints under sodium are of welded construction. The sealing is done using multiple mechanical seal arrangement equipped with a dedicated lubricating oil supply system.

(h) Capacity regulation by speed control: Capacity regulation in sodium pumps is achieved by varying the speed of the pump and not by throttling the discharge valve. Valves in sodium systems are restricted to minimise the possibility of leakage of sodium to the atmosphere from valve failure, thus improving system reliability, and avoiding energy loss due to valve throttling.

(i) Hydraulic characteristics: The arrangement consists of multiple pumps operating in parallel. The pump hydraulics is designed to achieve stable, drooping performance characteristic so that the pumps operate at the same duty point. Since capacity regulation is achieved by speed variation, the pump characteristics are designed to allow operation over a wide range of speeds (typically 15%–100% of rated speed) and over a range of flow rates (at a given speed). The latter criterion makes possible operation of the system in the event of (i) one pump trip or (ii) lower system resistance (e.g., duct or plenum failure).

(j) Other considerations: The overall design of the pump should facilitate easy assembly and disassembly for maintenance. Unlike conventional pumps, the sodium-wetted pump is to be cleaned from wet sodium using a special process (e.g., water-vapour/CO_2 process) before it is disassembled for maintenance. To prevent sodium accumulation in crevices/narrow regions, all sodium-wetted parts are designed to facilitate the complete drainage of sodium.

1.4 SUMMARY

The thermal energy generated by nuclear fission in the core of a fast reactor is transferred by liquid sodium to the conventional steam-water circuit to drive the turbogenerators and produce power. Either mechanical or electromagnetic pumps can be used to circulate liquid sodium in the primary and secondary circuits. Mechanical centrifugal pumps are preferred over electromagnetic pumps because of their proven design, high efficiency, rugged construction, and freedom from maintenance. Unlike conventional pumps used for water applications, sodium centrifugal pumps have unique design features and engineering requirements.

NOTES

1 Power density is a measure of the volumetric heat release rate in the core. Fast reactors are relatively compact to reduce fuel cost and neutron moderation. The typical power density in a sodium-cooled commercial fast breeder reactor is as high as 500 MW(thermal)/m^3. In comparison, the power densities in thermal reactors are approximately 100 MW(thermal/m^3 for pressurised water reactors, about 55MW(thermal)/m^3 for boiling water reactors, and about 19 MW(thermal)/m^3 for PHWRs.

2 The efficiency of electromagnetic pumps ranges from 5% to 45%, whereas that of centrifugal pumps of similar capacity is around 80% to 85%.

3 The Net Positive Suction Head (NPSH) is the total energy at the pump suction and is a measure of the cavitation performance of the pump. There are two types of NPSH, viz. NPSHA – Net Positive Suction Head Available and NPSHR – Net Positive Suction Head Required. NPSHA depends on the piping layout while NPSHR is controlled by the design of the pump impeller. To avoid cavitation, NPSHA > NPSHR.

REFERENCES

1. S.A. Bharadwaj, Indian Nuclear Power Programme – Past, Present and Future, *Sadhana*, 38: Part 5, 775–793, October 2013.
2. S. Glasstone and A. Sesonske, *Nuclear Reactor Engineering*, 3rd edition, C.B.S. Publishers and Distributors.
3. D.R. Nixon and C.C. Randall, Survey of Sodium Pump Technology, Report no. WCAP-2255 prepared for the New York Operations Office, U.S. Atomic Energy Commission, Under A.E.C. Contract AT (30-1)-3123, June 1963.

Chapter 2

Description of centrifugal sodium pumps

2.1 INTRODUCTION

Sodium-cooled fast reactors may be classified based on the design intent (e.g., research and development, technology demonstration, or power generation), into three categories: Experimental Fast Reactors, Demonstration/Prototype Fast Reactors, and Commercial Size Fast Reactors. Appendix 1 lists fast reactors designed/constructed/operated worldwide.

This chapter describes the primary and secondary pumps of typical reactors in each of the above categories to introduce the reader to the main features of a fast reactor pump, the overall size of the pump, the configurations used in different reactors, etc. This rather extended preamble will be helpful to the reader in imbibing the essence of the later chapters that discuss important topics such as design features, components fabrication, pump testing, instrumentation, and operating experience.

2.2 SODIUM PUMPS OF EXPERIMENTAL FAST REACTORS

Almost all experimental fast reactors, except for the EBR-II (now decommissioned) and the CEFR, were loop-type reactors. The primary sodium pumps of these reactors were located in separate shielded cells outside the reactor vessel and connected by piping to the reactor inlet and intermediate heat exchanger outlet. All the primary pumps in the experimental fast reactors are of the bottom suction concept, except for the pumps in RAPSODIE and FBTR, both of which are of the top suction type, with the pump taking suction from the pump tank (sump-type).

In bottom suction pumps, the hydraulic axial thrust acts downwards and complements the rotor weight, thus increasing the load on the top thrust bearing. On the other hand, in the top suction impeller the hydraulic axial thrust acts upwards, in opposition to the rotor weight, thus reducing the total load on the top thrust bearing. Balancing holes are often provided in the impeller to reduce the axial thrust.

DOI: 10.1201/9781003460350-2

2.2.1 BOR-60 (USSR) [1–3]

Figure 2.1 is the primary sodium pump of the BOR-60 reactor. It is a sump-type pump with liquid intake on one side of the tank. The rotating assembly, including the impeller and stationary parts, such as the diffuser and the

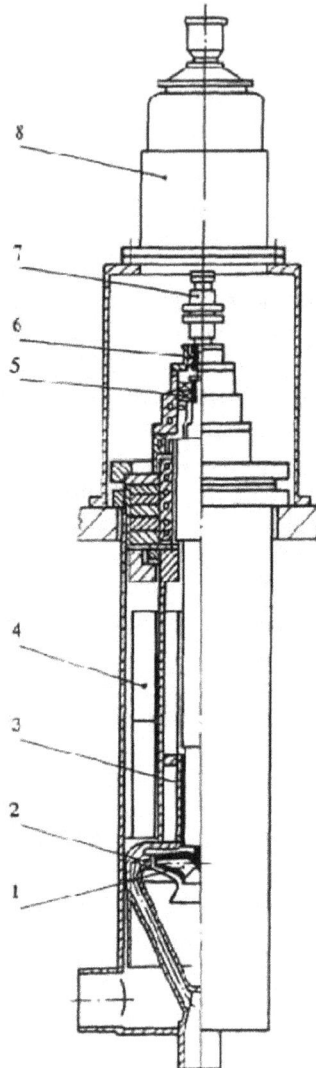

Figure 2.1 BOR-60 Primary Pump [3].1 – impeller; 2 – diffuser; 3 – radial bearing; 4– displacer; 5 – radial cum thrust bearing; 6 – shaft seal; 7 – coupling; 8 – motor. (Reproduced with kind permission of the International Atomic Energy Agency (IAEA)).

suction bell, are removable for maintenance. The liquid enters the pump tank through a side inlet, and a suction bell ensures smooth flow into the impeller inlet. Four pipes guide the sodium leaving the diffuser into a discharge chamber.

A hydrostatic bearing hardfaced with stellite guides the rotating assembly in sodium. The permitted fluctuation in level in the pump tank is 2 m, and gas entrainment into pump suction is prevented using a bell of 0.7 m height welded to the outer diameter of the pump diffuser.

Oil-lubricated double-row ball bearings provide radial and axial support to the top of the rotating assembly. Circulation of cooling oil through the bearing lubrication closed circuit is by a grooved sleeve keyed to the shaft. A mechanical seal located at the top of the shaft ensures the sealing of the pump cover gas. It consists of two pairs of sealing faces (graphite-nitrided steel pair) in a closed cavity filled with vacuum oil. The seals are designed for a pressure drop of 0.4 MPa and can be installed/removed as a unit for maintenance. A repair seal provides leak tightness during maintenance of the top bearing and seal. A standard motor with stepless speed control drives the pump.

The secondary pump is similar in construction to the primary pump but differs in size and the absence of shielding.

2.2.2 Fast Breeder Test Reactor (FBTR) (India) [4]

2.2.2.1 Primary pump

The FBTR primary pump (Figure 2.2), adapted from the advanced version of the RAPSODIE/FORTISSIMO pumps, differs from the original RAPSODIE pump in the sizing of hydraulic parts and the shaft/bearing system while retaining other basic features. The FBTR pump has a float-type non-return valve (NRV) after the pump discharge to prevent reverse flow through the stopped pump. A unique feature of this NRV is that the (movable) float of the valve is in neutral equilibrium with sodium (density of the valve is the same as that of sodium) under normal operating conditions, thereby avoiding sudden closure of the discharge pipe and the consequent water hammer effect. The hydrostatic bearing receives feed from the pump discharge through a drilled hole in the shaft. A strainer provided at the entry to the hole prevents clogging of the throttling orifice in the rotating journal of the hydrostatic bearing. The bearing mating surfaces are hardfaced using Colmonoy, a nickel-based hard-facing alloy. In the mechanical seal, the stationary ring, made of steel, is hardfaced with Stellite, which engages with the rotating ring made of graphite. The arrangement of the upper bearing facilitates the removal of the bearings and seal cartridge as a single unit.

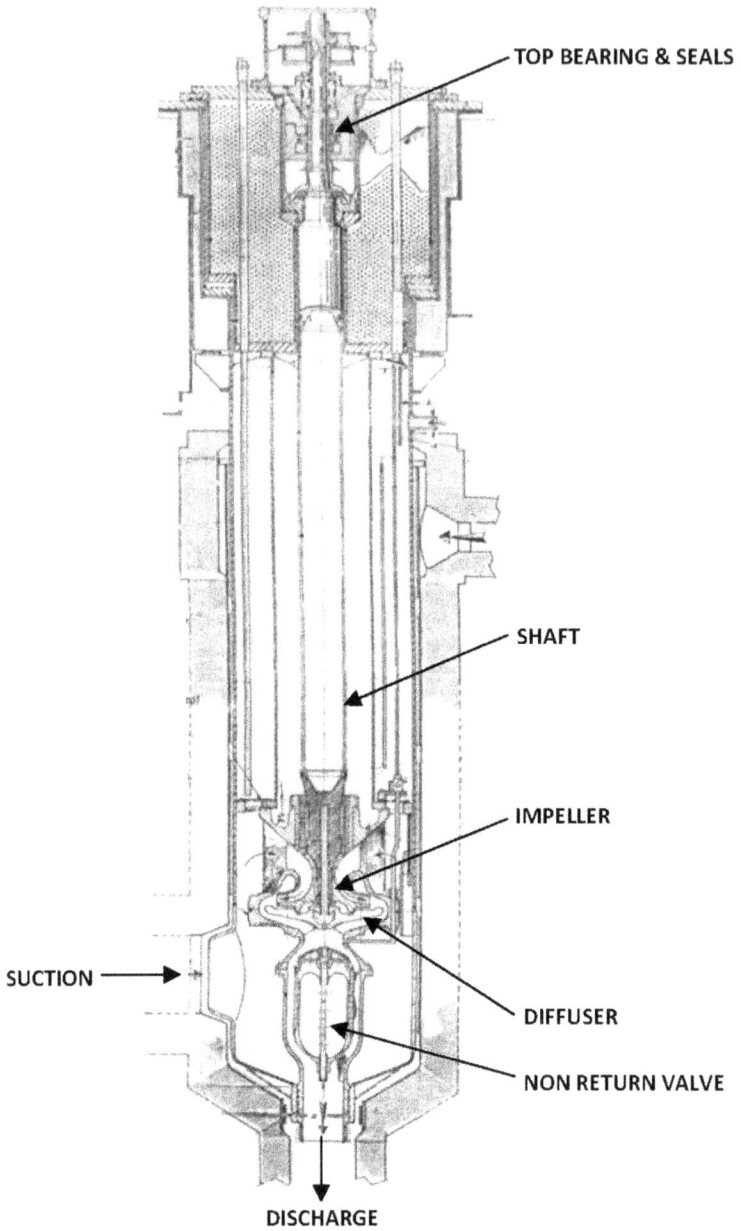

Figure 2.2 FBTR primary pump.

2.2.2.2 Secondary pump

The secondary sodium pump of FBTR (Figure 2.3) is similar to the primary pump except that it is shorter in height (2340 mm) compared to the primary pump (~5000 mm) due to the absence of radiation shielding. The pump is

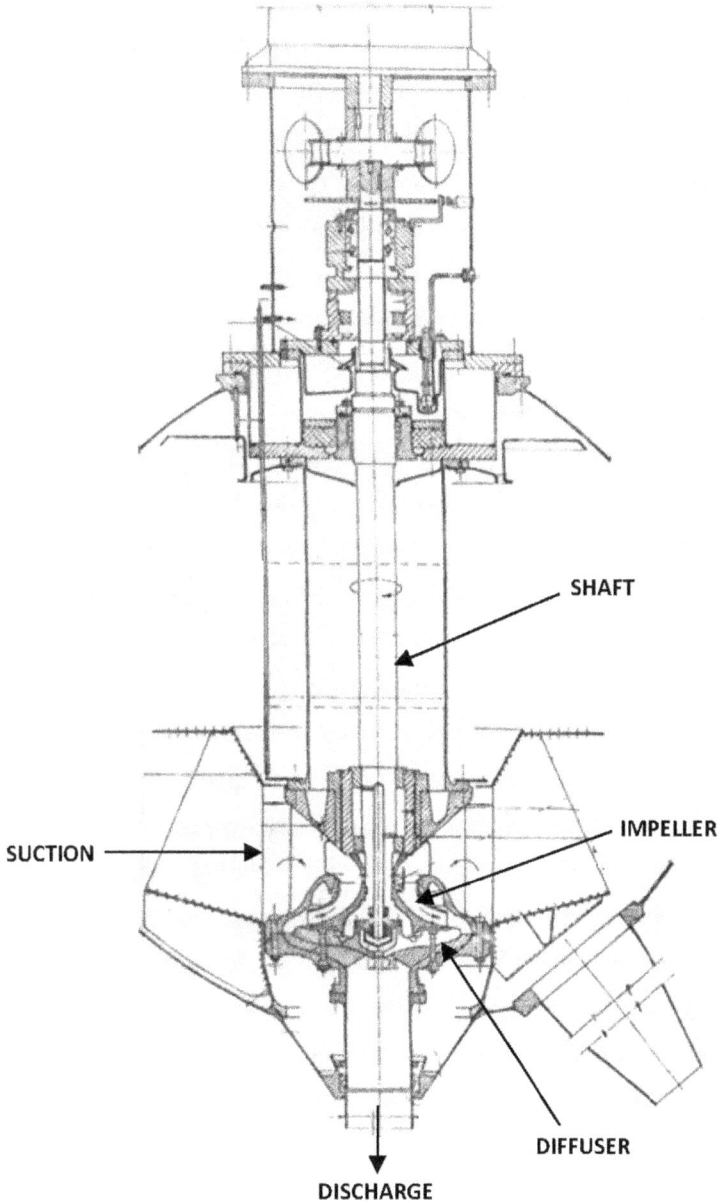

SHAFT

IMPELLER

SUCTION

DIFFUSER

DISCHARGE

Figure 2.3 FBTR secondary pump.

housed in a spherical vessel called an expansion tank, and has a top suction impeller and vaned diffuser. There is no non-return valve at the pump outlet because the two secondary circuits operate independently. As in the primary pump, the rotating assembly is supported in sodium by a hydrostatic bearing and outside the cover gas by taper roller bearings.

2.2.3 Experimental Breeder Reactor (EBR-II) (USA) [5]

The EBR-II (U.S.) pumps are of vertical construction with rectangular (biological) shield plugs. The impeller is of the bottom suction type that draws sodium directly from the primary reactor tank. The pump assembly consisting of shield plug, shaft, baffles, volute casing, and impeller is removable from the primary tank for maintenance. The rotating assembly is supported in sodium by a four-pocketed hydrostatic bearing hardfaced with Colmonoy.

The primary system consists of two primary pumps operating in parallel. The primary pump (Figure 2.4) assembly consists of the impeller and diffuser, 3.2 m long shaft, baffle assembly, shield plug, and a hermetically sealed drive motor.

The material of construction is austenitic stainless steel types 304 and 316. The hermetically sealed motor enclosure is filled with argon cover gas communicated to the pump tank cover gas. The gas in the motor enclosure is circulated through argon to air heat exchanger to maintain its temperature. Two bearings support the motor shaft: the top oil-lubricated angular contact ball bearing takes both radial and thrust loads, while the bottom roller bearing is grease-lubricated. The shaft is of composite geometry (partly hollow and partly solid) and is connected to the motor shaft by a rigid coupling. The top of the pump shaft passes through a shield section, about 1.95 m thick, before it is coupled with the motor. Four horizontal baffles are provided around the shaft just below the shield section to limit the heat transferred to the top of the rotating assembly. Two labyrinth seals, one at the top of the baffles assembly and the other below the drive motor, limit the flow of radioactive cover gas and sodium vapour to the motor enclosure. The motor housing is purged with argon gas at a flow rate of 5 to 7 ft^3/h, which further limits the convection of sodium vapour. The pump flow rate is regulated by varying the speed of the pump between 20% and 100% of the rated speed. A variable slip coupling connects the motor with the generator that drives the motor. The coupling varies the speed of the generator. The secondary pump is AC linear induction type electromagnetic pump.

2.2.4 Enrico Fermi Atomic Power Plant (EFAPP also known as Fermi) (USA) [6, 7]

The EFAPP (aka Fermi) is a 430 MWt reactor with three primary and secondary pumps. The rotating assembly of the primary pump (Figure 2.5) is supported

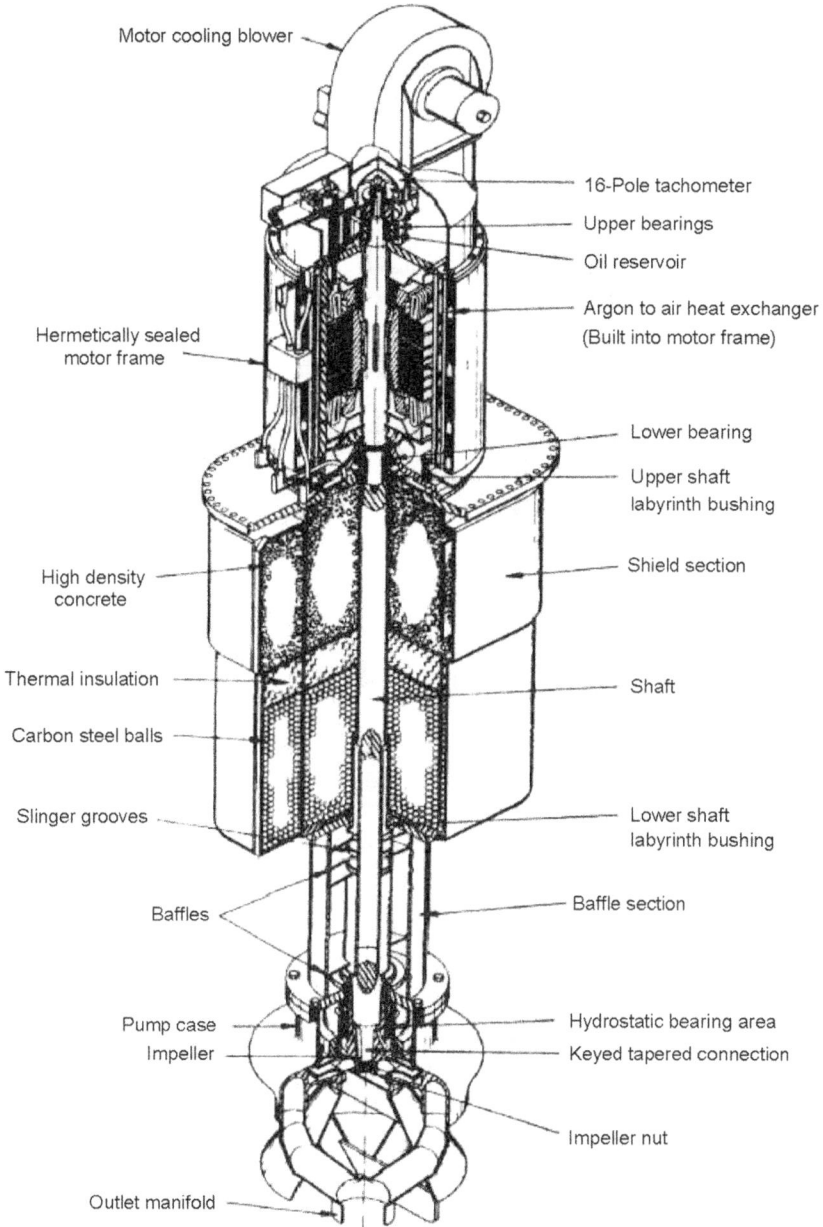

Figure 2.4 EBR-II Primary Pump [5]. (Reproduced with kind permission of Argonne National Laboratory, managed and operated by UChicago, Argonne, LLC, for the U.S. Department of Energy under Contract No. DE-AC02-06CH11357).

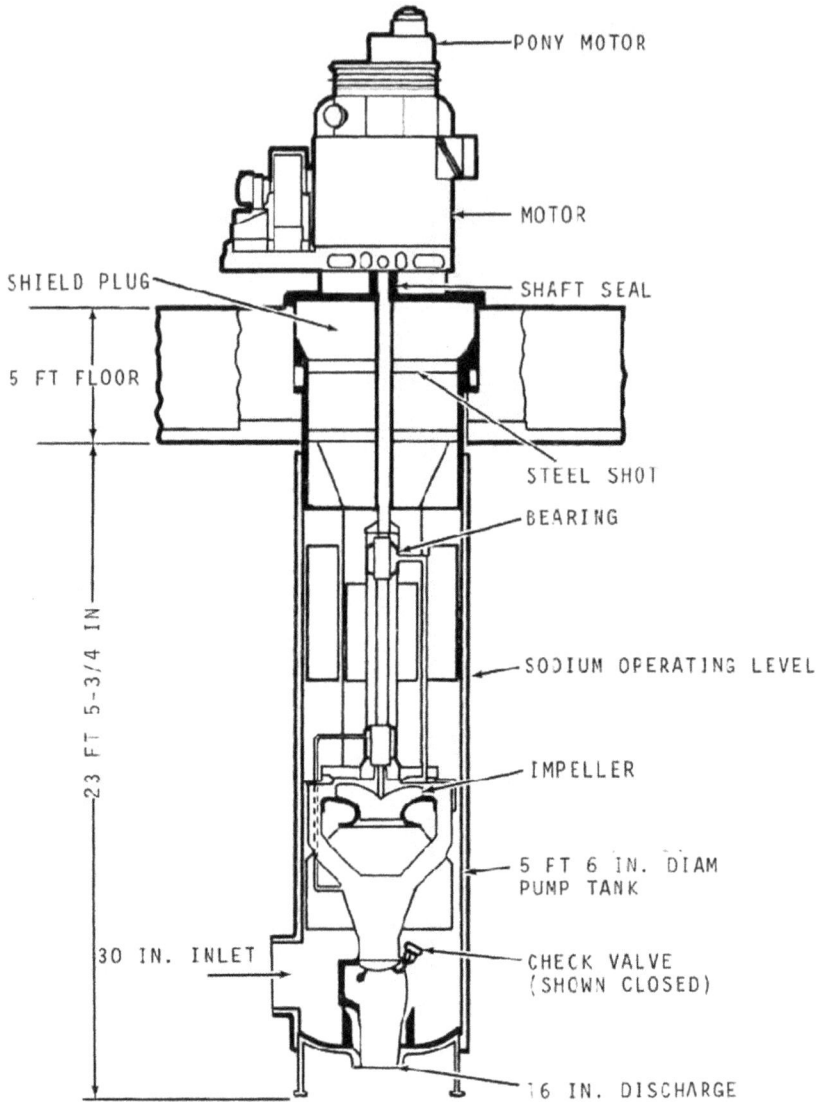

Figure 2.5 EFAPP primary pump [6]. (Reproduced from, W. Babcock, State of Technology Study – Pumps: Experience with high temperature sodium pumps in nuclear reactor service and their application to FFTF, AEC Research and Development Report no. BNWL-1049 UC-80, Reactor Technology, Dec. 1969BNWL – Batelle Northwest Laboratories, Richland, Washington, USA).

by two hydrostatic bearings for a shaft length of 5.4 m. The impeller is of single stage, bottom suction type, and the impeller suction is in communication with the sump/pump tank (sump suction concept). The system piping connects to the pump tank at the side. A draw bolt couples the pump and motor shafts with a shaft seal at the interface. The pump has a check valve at the discharge which facilitates the isolation of the stopped pump. A rigid coupling couples the pump and motor shafts.

Unlike the primary pump, the secondary pump (Figure 2.6) rotating assembly is supported inside sodium by a single hydrostatic bearing with the other bearing (radial and thrust bearing) located outside the pump tank.

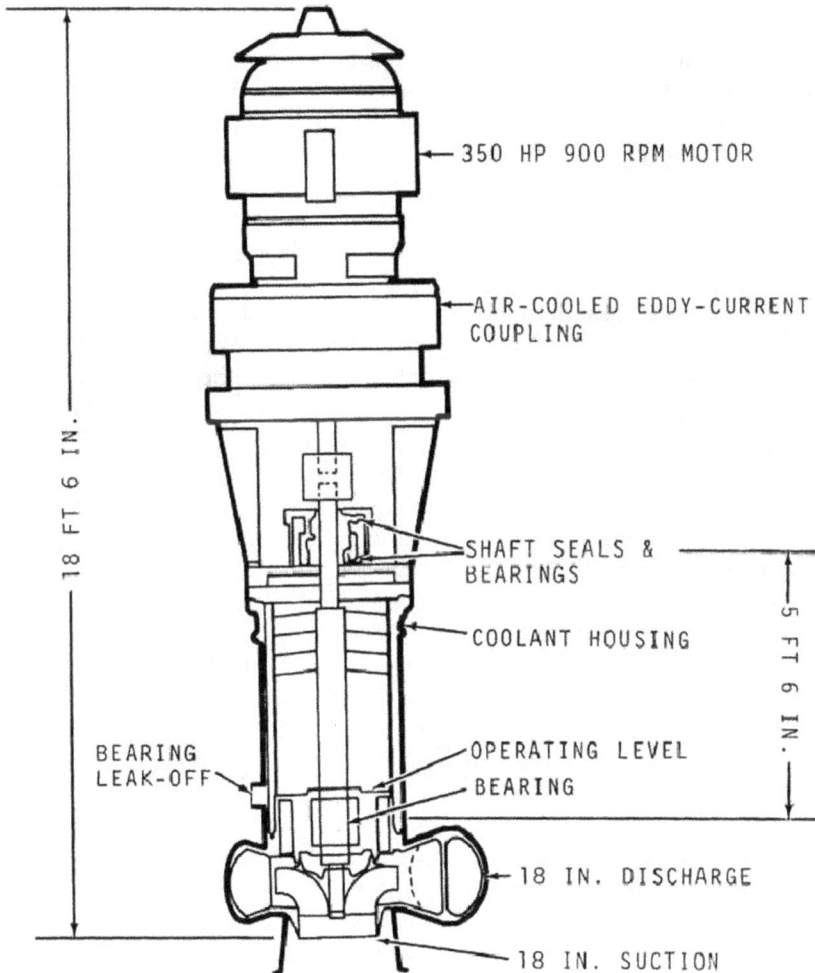

Figure 2.6 EFAPP secondary pump [6]. (Reproduced from W. Babcock, State of Technology Study – Pumps: Experience with high temperature sodium pumps in nuclear reactor service and their application to FFTF, AEC Research and Development Report no. BNWL-1049 UC-80, Reactor Technology, Dec. 1969BNWL – Batelle Northwest Laboratories, Richland, Washington, USA).

2.2.5 Fast Flux Test Facility (FFTF) (USA) [8, 9]

The primary sodium pump (Figure 2.7) in the Fast Flux Test Facility was one of the largest centrifugal sodium pumps operated in the USA until the mid-nineties. The three primary pumps are vertical, free-surface, single-stage designs driven by wound rotor electric motors.

The primary pump is in the hot leg inside the pump tank that is supported from the top at the operating floor level. A guard vessel is provided around the pump tank, and the interspace is accessible for in-service inspection. The principal material of construction of the pump is austenitic stainless steel AISI 304.

Low-pressure coolant enters the pump tank through the suction nozzle and is directed to the impeller by the suction elbow. The gap (~19 mm) between the suction nozzle and the suction elbow ensures favourable suction conditions similar to that with a free liquid surface. Sodium leaving the impeller is directed by the diffuser assembly to the pump discharge nozzle. The clearance between the suction nozzle of the pump tank and the suction elbow of the internals facilitates the removal of the pump assembly from the tank in the event of a maintenance requirement.

As the pump is located in the hot leg, it has to sustain the high operating temperature of 565°C as well as severe thermal transients. The hydraulic assembly is designed to ensure that only non-pressure bearing walls experience the complete transient. The pressure bearing parts, such as the pump casing is subjected to attenuated thermal transients by mixing the constant temperature liquid in the pump with the transient liquid passing through the pump. At the bottom of the diffuser about 0.3% of the flow enters the cavity formed by the casing and the diffuser, mixes thoroughly, and flows upward into the hydrostatic bearing supply reservoir through channels in the annular space between the casing and the diffuser wall. The high-pressure sodium then enters the hydrostatic bearing pockets.

The low velocity in the casing causes the settling out of any particles in the flow (of the order of 1–2 mils (25 μm–50 μm) in size), thereby preventing choking of the narrow clearance of the hydrostatic bearing.

At the operating speed of 900 rpm the vortex formed in the sodium in the annular space between the pump shaft and the support cylinder causes the space to be pumped dry, thereby reducing the viscous friction between the rotating shafts. The resulting reduced added mass also increases the critical speed of the rotating assembly.

The primary pump shaft consists of a central hollow portion welded to solid forgings at either end to optimise the stiffness-to-mass ratio. The hollow part of the shaft is evacuated to avoid internal cellular convection, possible shaft bowing, and unbalance. The total length of the primary shaft is 7.2 m.

The rotating assembly is supported in sodium by a four-pocket hydrostatic bearing of size 300 mm ×300 mm with a radial clearance of 280 μm. Standard hydrodynamic pivoted pad-type radial bearing and Kingsbury-type thrust bearing with oil jacking provide support at the top of the rotating

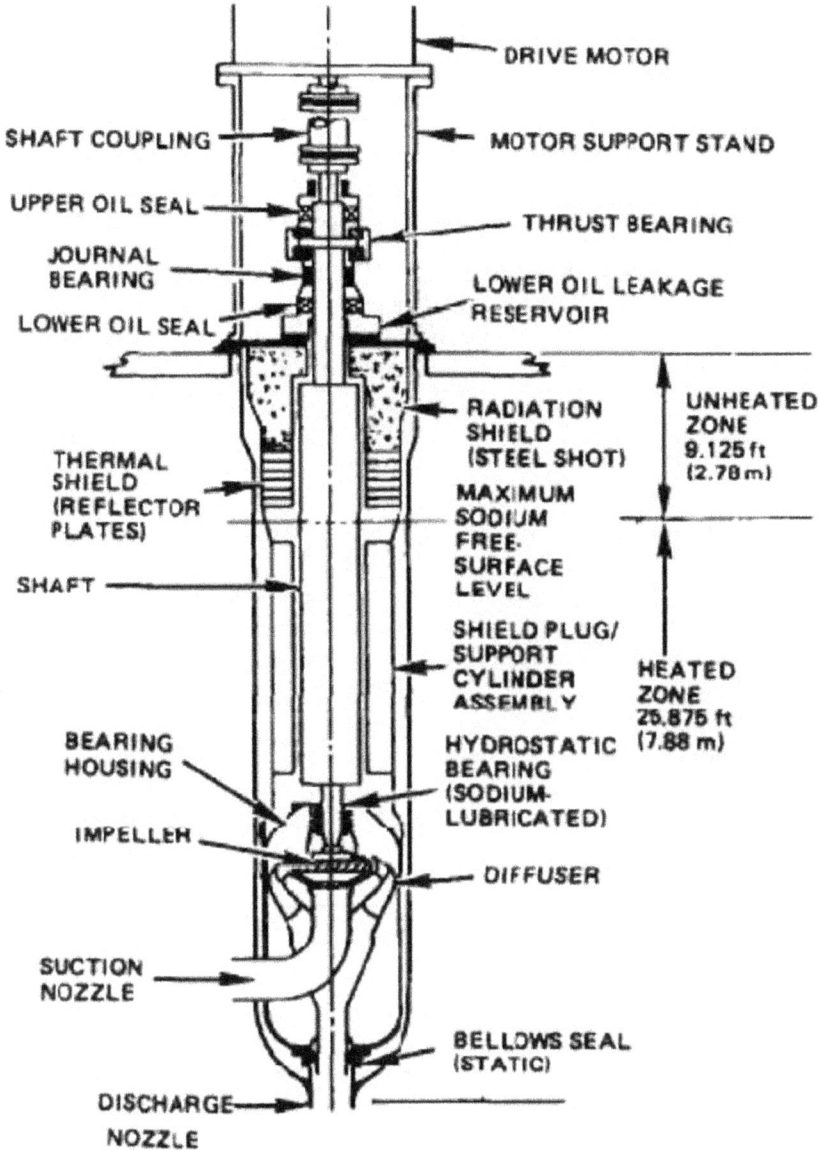

Figure 2.7 FFTF primary pump [8]. (Reproduced from R. W Atz, M. J Tessier, High Temperature Testing of a Sodium Pump, Presented at ASME Winter Annual Meeting, December 10–15, 1978).

assembly. The first lateral critical speed of the rotating assembly is 1.25 times the design speed. Two mechanical seals, one on either side of the bearing assembly, seal the pump cover gas. A common oil lubrication system is used for the bearings and seals. Thermal baffles in the cover gas space reduce the thermal load to the top bearings and seals, and the shield plug at the top of the pump provides biological shielding from radiation streaming.

The secondary sodium pumps of FFTF are hydraulically similar in design to the primary pumps. The main difference is that the secondary sodium pump is shorter (shaft length 4.2 m) with no biological shielding.

2.2.6 Hallam Nuclear Power Facility (HNPF)[1] (USA) [6, 7]

The Hallam Nuclear Power Facility (HNPF) was a prototype sodium-cooled, graphite-moderated thermal reactor. This loop-type reactor of 90 MWe capacity was one of the earliest sodium-cooled reactors and started power generation in 1964. The primary (3 no.) and secondary coolant pumps (3 no.) of this facility are discussed here because they are representative of the earliest sodium centrifugal pumps of piped suction design (in contrast to the EFAPP pump which was one of the earliest pumps of the sump suction design).

The hydraulics of both primary and secondary sodium pumps were identical, with both pumps provided with single-stage, bottom suction impeller, and vaned diffuser. The pump suction was connected to system piping (piped suction) and the pump discharge was connected to the pump tank. A sealing ring separated the pump tank, high-pressure side from the cover gas low-pressure side, and the liquid overflowing from the pump tank was piped back to the pump suction. The rotor assembly was supported in sodium by a hydrostatic bearing and by at the top by oil-lubricated ball bearings. Mechanical seals provided leak tightness of pump cover gas and prevented oil leaks into sodium or the atmosphere. A sketch of the primary pump is shown in Figure 2.8.

2.3 SODIUM PUMPS OF DEMONSTRATION/PROTOTYPE FAST REACTORS

Pumps used in primary and secondary coolant circuits of large reactors adopt vertical pump designs similar to that used for the experimental reactors. However, their duty calls for very large flow rates, although the head requirements are only marginally higher. On account of modest suction side pressure, particularly in pool-type reactors, and high capacity at duty point, the operating speeds are considerably lower – in large pumps such as the ones of the SuperPhénix reactor, they are less than 500 rpm. The low operating speed is necessary for cavitation-free operation over a long service life of 40–60 years.

2.3.1 BN-350 (USSR) [1–3]

The BN 350 primary and secondary pumps are similar and differ only in the impeller and diffuser hydraulics. The delivery header is a part of the tank in these pumps, unlike that in the BOR-60 pumps.

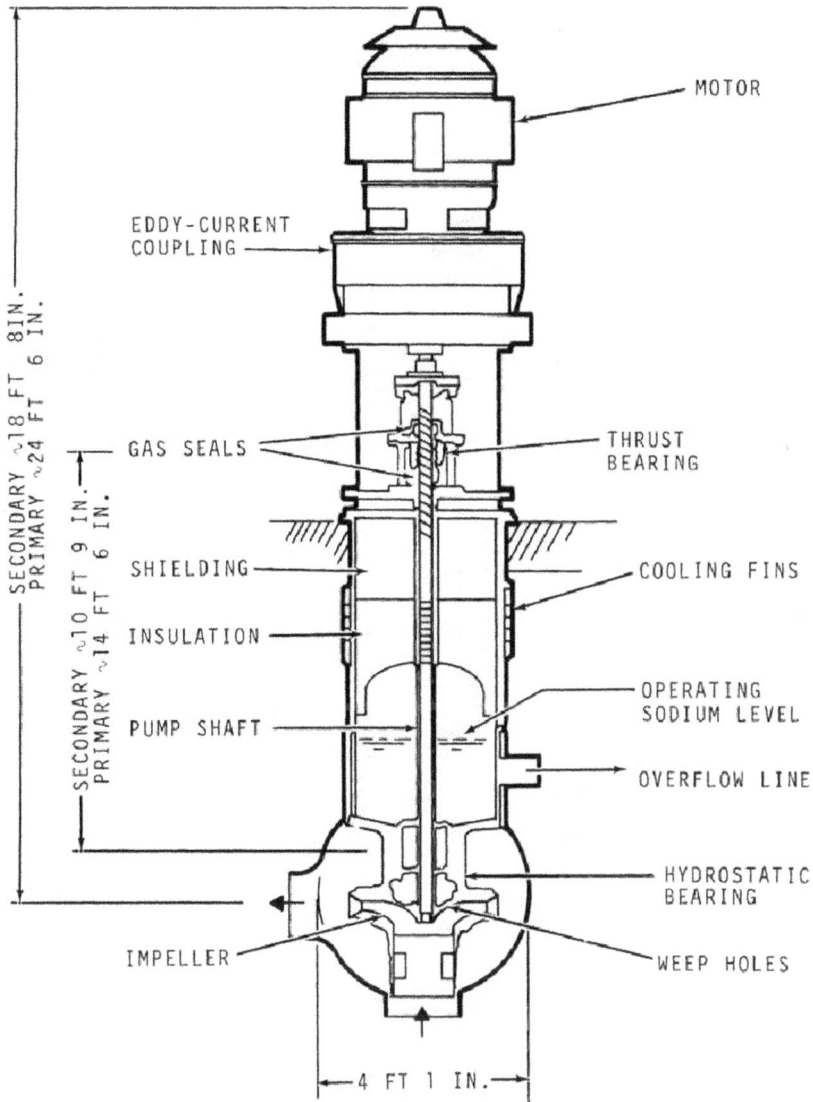

Figure 2.8 Primary pump of HNPF [6]. (Reproduced from W. Babcock, State of Technology Study – Pumps: Experience with high temperature sodium pumps in nuclear reactor service and their application to FFTF, AEC Research and Development Report no. BNWL-1049 UC-80, Reactor Technology, Dec. 1969BNWL – Batelle Northwest Laboratories, Richland, Washington, USA).

The rotating assembly of the BN 350 primary pump (Figure 2.9) is supported outside the pump tank top flange by an upper radial and thrust bearing and a lower radial bearing. The working faces of both bearings are babbited. The radial bearings have non-split sleeves, while the thrust bearing

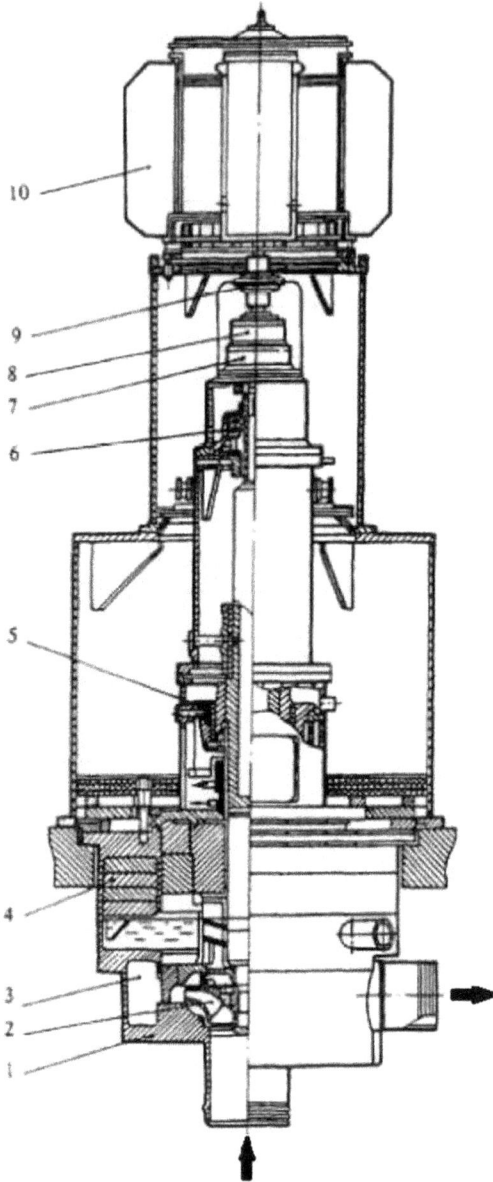

Figure 2.9 Primary pump of BN-350 [3]. I – casing; 2 – impeller; 3 – high-pressure header; 4 – biological shield; 5 – radial bearing; 6 – radial cum thrust bearing; 7 – repair seal; 8 – face seal; 9 – coupling; 10 – motor. (Reproduced with kind permission of the International Atomic Energy Agency (IAEA)).

is a self-aligning pad bearing (Mitchell bearing). External closed-circuit oil lubrication is used for the bearings. However, unlike the bearing in the BOR-60 pump, the oil delivered to the bearings is independent of the pump speed. The rotating assembly is of cantilever construction being unsupported in sodium with the overhang of the impeller from the lowest bearing being 2 m.

Sodium enters the bottom suction impeller through the inlet pipe at the bottom of the tank, and high-pressure sodium is discharged through the side outlet. The impeller is provided with balancing holes that minimise the axial thrust on the rotating assembly. Gas entrainment into pump suction is prevented by ensuring sufficient submergence in the pump tank by diverting a portion of high-pressure discharge into the pump tank through a slotted seal on the shaft; the excess liquid is drained to the main circuit through a pipe. Sodium leaking through the gap between the removable internals and the stationary pump tank is diverted back to pump suction through a bypass line connected to a tank in which the level is regulated. This, in turn, maintains the sodium level in the pump tank under various operating conditions. The shaft is of hollow construction in the central portion, and a mechanical seal (similar to that in the BOR-60 pump with a pressure drop capacity of 0.4 MPa) is used to seal the cover gas in the pump tank. A repair seal made of rubber is provided below the mechanical seal to maintain the leak tightness of cover gas in the pump in the event of replacement of the mechanical seal. A cooler is provided above the pump tank top flange to reduce the temperature at the bottom bearing location.

The pump is driven by a squirrel cage induction motor with two speeds, viz., 250 rpm and 1000 rpm.

2.3.2 BN-600 (USSR) [1–3]

2.3.2.1 Primary pump

In the BN-600 reactor primary sodium pump (Figure 2.10), the opposing requirements of a modest net positive suction head and a compact design are satisfied using a double suction impeller. The sodium is guided to either half of the impeller axially through channels in upper and lower volute passages and delivered from the impeller through the guiding duct and vertical channels in the lower volute. The double suction impeller permits a significantly high operating speed of 970 rpm, thereby minimising the pump lateral dimension. The castings for the pump hydraulics are made from steel 10X18H12M3Л.

The rotating assembly is supported in sodium by a radial hydrostatic bearing fed from the impeller outlet. The bearing is double-split throttling type, and the surface of the stationary bearing sleeve and the mating portion on the shaft are stellited. Throttling of sodium entering the bearing chambers occurs in the gap between the rotating shaft and the stationary bearing sleeve. The liquid exiting the hydrostatic bearing flows into the cover gas

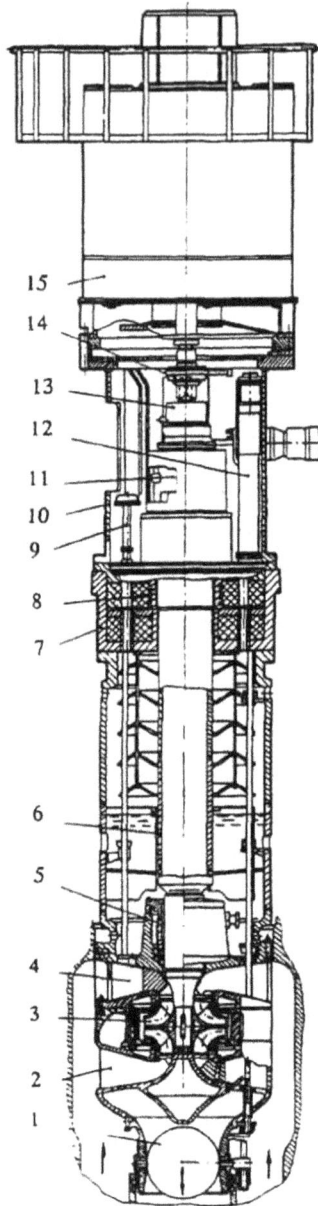

Figure 2.10 BN-600 Primary Pump [3]. 1 – check valve; 2 – lower scroll; 3 – impeller; 4 – upper scroll; 5 – hydrostatic bearing; 6 – shaft; 7 – cover; 8 – cooler; 9 – Level gauge; 10 – motor base; 11 – face seal; 12 – check valve drive; 13 – radial-thrust bearing; 14– coupling; 15 – motor. (Reproduced with kind permission of the International Atomic Energy Agency (IAEA)).

region in the pump tank and is redirected back to the pump inlet through the annular gap between the removable pump internals and the pump tank and four throttling orifices. This arrangement prevents any undue level rise in the cover gas space of the pump tank while ensuring the minimum submergence required to prevent gas entrainment.

The pump shaft, hollow in the middle portion, is made of steel 10X18H9 and is constructed from six segments welded together to maximise the stiffness to weight ratio. The material of construction of all other parts is X18H9.

The length of the shaft is 7.6 m and its maximum diameter is 0.68 m. A toothed coupling of variable stiffness connects the pump shaft to the motor shaft.

Thermal baffles are provided in the cover gas space to reduce the heat transferred to the top cover of the pump. The top cover provides shielding from radiation streaming and consists of steel and graphite plates of a total 1000 mm thickness (500 mm steel + 500 mm graphite). A water cooler in the top cover keeps the temperature in the cover within limits.

The upper radial bearings are identical to that in the BN 350 pump. The thrust bearing, however, is of Kingsbury-type and uniform load distribution on the load pads is ensured by a set of rings made of spring steel. The upper bearings are lubricated by a dedicated external system consisting of pumps, filters, cooler, oil tank, and valve mounted as a single oil-block assembly on a shared bed plate. Turbine oil is used as the lubricant.

The cover gas is sealed using mechanical seals as in the BOR-60 and BN 350 primary pumps. The seal is removable as a unit and consists of two friction pairs of graphite-nitrided seal faces enclosed in a chamber filled with vacuum oil to form a hydraulic lock. The rotating rings of the friction pairs are freely movable axially and radially relative to the shaft, thus ensuring self-aligning of the stationary rings relative to the seal case. The cover gas seal is located below the upper bearing and this arrangement prevents oil ingress into the pump. A catchpot below the seal prevents such oil ingress under emergency conditions. The seals are designed for a pressure of 0.4 MPa.

A maintenance seal (repair or standby seal) is provided, below the mechanical seal, to seal the pump cover gas during the replacement of the mechanical seals. The sealing parts of the repair seals are made of fluoro plastic rings and are fixed in a movable flange. The repair seal is tightened by pressing the sealing ring to the shaft by pressurising the working bellows.

A flap-type check valve is provided in the pump discharge nozzle to prevent reverse flow of sodium in the event of a pump trip. The drive of the check valve is of hydraulic type and is actuated by signals from the automatic control system.

The pump is driven by a controllable synchronous motor coupled with asynchronous rectifier driver with phase rotor, and a toothed coupling is used to connect the motor to the pump.

2.3.2.2 Secondary pump

The secondary pump (Figure 2.11) in each loop of the BN-600 reactor is located in a tank. The tank has external electric heaters for preheating before sodium filling and maintaining the temperature in the range of 200°C–250°C. The pump has higher available NPSH (the cover gas pressure in the secondary pump is 0.2 MPa) when compared to the primary pump, and is therefore provided with a single suction impeller. The flow is discharged through a double spiral volute in the pump tank.

Many of the components in the secondary pump are identical to that in the primary pump. These include shaft seal, repair seal, top radial cum axial bearing and pump coupling. The hydrostatic bearing is smaller in diameter (350 mm) than that in the primary pump, but the design is similar. The impeller wearing rings in the primary and secondary pumps are of labyrinth type. The gaps in the wearing ring locations are 1–2 mm to facilitate easy assembly and ensure contact-free operation under unexpected temperature differential or mechanical deformation of parts.

2.3.3 Phenix (France) [1, 3, 10, 11]

The pumps in the Phenix reactor are one of the earliest large reactor pumps. Three pumps are used in each of the primary and secondary circuits.

The primary pump (Figure 2.12) is a vertical, centrifugal pump with a free level of sodium. The hydraulic design is primarily extrapolated from the RAPSODIE/FORTISSIMO primary pump and consists of a top suction impeller and diffuser. A non-return valve, similar to that in the RAPSODIE pump, is provided in the discharge pipe and prevents reverse flow through the stationary pump when one pump in the loop is not in operation. The total length of the motor and pump assembly up to the delivery nozzle is 17 m. The pump rotor assembly, over 5 m in length, is supported at the top by a double-row roller bearing and inside sodium by a hydrostatic bearing. High-pressure sodium is fed to the hydrostatic bearing from the pump discharge.

The diameter of hydrostatic bearing is 320 mm, and the radial clearance is 0.5 mm. The bearing rigidity was validated to be sufficient to limit the shaft displacement within the bearing in the range of 20% of the radial clearance using water tests. Testing the pump at approximately 650 rpm speed showed good operating performance of the bearing. The maximum pump operating speed is 960 rpm; however, the operating point in the reactor was achieved at a lower speed of 820 rpm due to lower than estimated head loss in the circuit (the secondary pump achieved the required flow rate at a speed of 800 rpm). A mechanical double-face seal with oil lock seals the shaft. A repair seal is provided to maintain leak tightness of the pump cover gas in the event of removal of the upper parts of the pump (seal, bearing etc.) for maintenance/repair. In such instances, the bearing/seal cooling oil circuit is depressurised, and the repair seal is engaged before the removal of the parts.

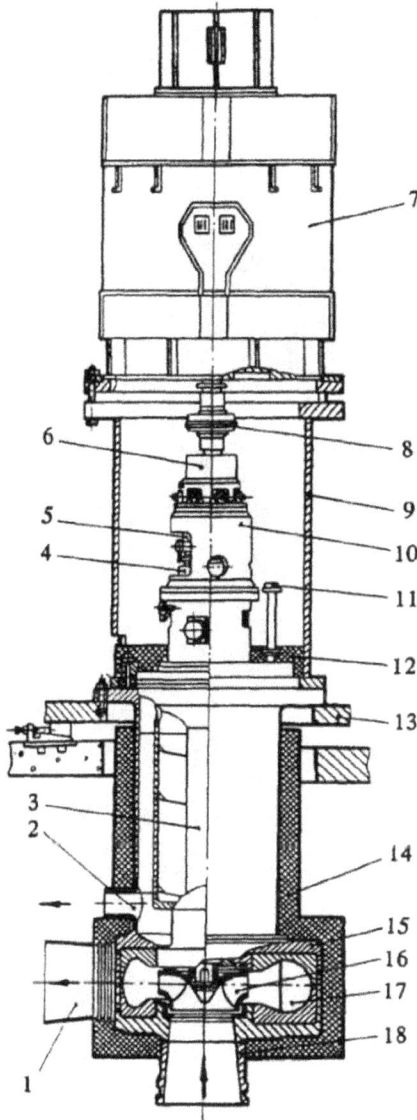

Figure 2.11 BN-600 secondary pump [3]. 1– discharge nozzle; 2 – leakage flow at discharge; 3 – shaft; 4 – repair seal; 5 – face seal; 6 – radial-thrust bearing; 7 – motor; 8 – coupling; 9 – base; 10 – removable part; 11 – level gauge; 12 – thermal insulation; 13 – support plate; 14 – electric heater and thermal insulation; 15 – tank; 16 – impeller; 17 – scroll; 18 – suction nozzle. (Reproduced with kind permission of the International Atomic Energy Agency (IAEA)).

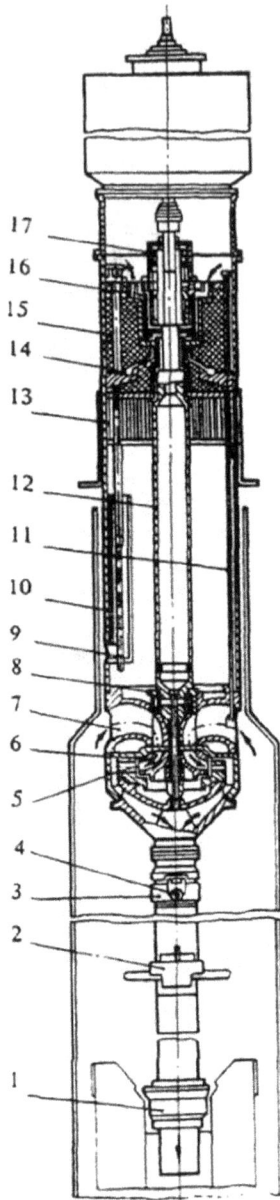

Figure 2.12 Phenix primary pump [3]. 1 – lower seal; 2 – flow meter; 3 – upper seal; 4 – hinge; 5 – guiding duct; 6 – impeller; 7 – inlet scroll; 8 – hydrostatic bearing; 9 – level gauge; 10 – pump casing; 11 – thermocouple; 12 – shaft; 13 – thermal insulation; 14 – catch pot for leaked oil; 15 – biological shield; 16 – air cooling shroud; 17 – bearing shaft seal unit. (Reproduced with kind permission of the International Atomic Energy Agency (IAEA)).

The shield plug at the top of the pump is filled with cast iron pellets to provide the biological shield (15). A double gasket on the top flange of the plug seals the radioactive gas in the pump cavity from the atmosphere. The gasket is made from heat-resistant grade of rubber. A layer of insulation (13) is provided below the shield plug to reduce the heat flow towards the top assemblies of the pump.

A flywheel is provided to maintain the required flow rate to the core for three minutes during a loss of power supply.

The design improvements in this second-generation pump with respect to the earlier design used in RAPSODIE are as follows:

(a) The discharge nozzle of the pump is connected to the grid plate pipe through a hinge joint (also known as articulated sleeve). This design eliminates misalignment and displacement of the pump relative to the grid plate piping while absorbing the lateral movement of the grid plate piping and achieving good leak tightness.

(b) The upper pump support is designed to properly position the pump on the roof slab without unduly stressing the joint between the pump and the roof slab because of radial thermal expansion, during reactor operation, of the pipe receptacle connecting the pump discharge nozzle to the grid plate.

(c) The pump is driven by a synchronous motor with frequency controlled variable speed system that permits speed regulation from 250 rpm to 975 rpm. In addition, a battery-driven auxiliary drive motor fed from a low-voltage auxiliary grid through a rectifier is also provided. In the case of a prolonged power supply interruption, the flywheel inertia achieves a gradual reduction of the pump speed up to 100 rpm, wherein the auxiliary drive motor is coupled to the pump. In this mode, the auxiliary drive permits operation for one hour.

(d) A brake applied to the flywheel rim prevents pump operation in turbine mode should the check valve (non-return valve) fail to close.

The secondary sodium pumps are built into expansion tanks and are similar in design to the primary pumps except for the absence of a flywheel, non-return valve, and biological shielding.

2.3.4 Prototype Fast Reactor (PFR) (UK) [12–14]

2.3.4.1 Primary pump

The designs for the PFR primary and secondary sodium pumps was based on extensive work done on an experimental sodium pump between 1964 and 1968 in the Risley Engineering Laboratory, UK which provided about 8,000 h of invaluable operating experience.

There was no experience with mechanical sodium pumps in the UK, unlike in France, the erstwhile USSR, and the US, where mechanical pumps were used in experimental fast reactors. The UK experimental pump was a medium-capacity pump with flow rate 1350 m³/h, unlike other small reactor pumps, which had capacities well below 1000 m³/h. It was of single-stage bottom suction with a diffuser and casing, all of which could be removed as a unit for maintenance. The pump was tested in water and sodium, and these tests were instrumental in deciding the cavitation criterion of no visible cavitation for the primary pump.

The primary circuit consists of three pumps that operate in parallel and circulate sodium from the pool through the reactor core. The outlet of the three pumps connects to the diagrid through 12 high-pressure pipes. The primary pump is of double suction type to satisfy the cavitation criterion of no visible cavitation, and the design was finalised after detailed tests in water on scaled models. The pump has a four-segment volute arrangement with four separate outlet pipes, and this arrangement improves the flexibility of the piping between the pump and the non-return isolation valve in the discharge pipe, thus minimising the load on the pump casing.

The PFR primary pump (Figure 2.13) is, like the BN-600 pump, a shallow submerged pump. The choice of a double suction impeller is appropriate considering the limited NPSH available and in eliminating the hydraulic thrust that would have otherwise occurred. The method of attaching the impeller to the shaft addresses the conflicting requirements of obtaining a satisfactory fit to transmit the torque without fretting wear at the impeller hub and the need to ensure easy assembly and disassembly of the impeller.

The shaft of the primary pump consists of a central tubular portion welded at its top and bottom ends to solid forgings suitably profiled to mount the bearings and impeller, respectively. The rotating assembly is guided in sodium by a hydrostatic bearing fed from the pump outlet while it is supported at the top by integral oil-lubricated, hydrostatic, annular slot sleeve and hydrodynamic Raleigh thrust step bearing. The pump cover gas argon is sealed from the atmosphere by rubbing face shaft seal. In this arrangement, oil flow from the bottom of the annular bearing provides oil at 0.85 kg/cm² pressure to back up the shaft face seal. Hence a single oil system supplies both the seal and the top bearing. The span between the top and bottom bearing of 4.55 m (15 feet) ensures acceptable temperature at the top bearing while maintaining adequate submergence of the impeller.

The pump shaft is connected to the motor shaft through an intermediate shaft. A flywheel on the motor shaft ensures a minimum coastdown time of 8 secs from full speed to half speed.

The flywheel is provided with a protective forged rim that is shrunk fit on it to prevent crack propagation and ejection of fragments in the event of disintegration of the flywheel.

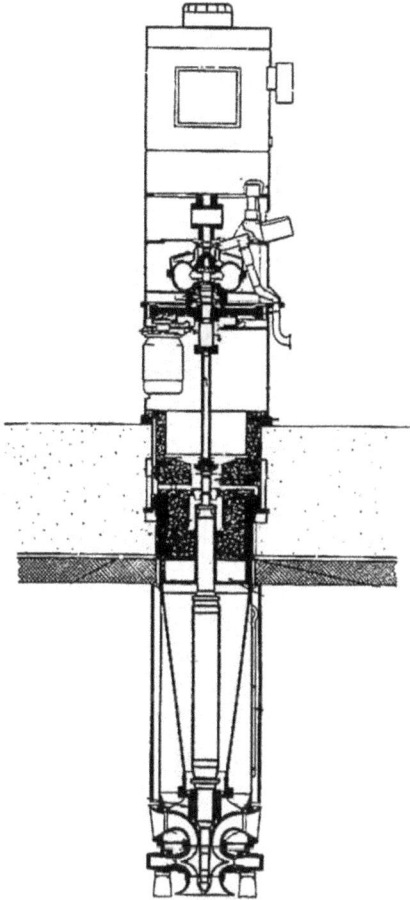

Figure 2.13 PFR primary pump [12]. (Reproduced with kind permission of the International Atomic Energy Agency (IAEA)).

A fixed-speed main induction motor drives the primary pump through a fluid coupling that facilitates speed regulation from 20% to 100% of nominal speed.

A valve is provided on each pump's discharge side to prevent bypassing the core flow due to reverse flow through the stationary pump when only two out of three pumps are in operation. This arrangement facilitates reactor operation with one pump tripped. Details are given in Chapter 3.

2.3.4.2 Secondary pump

A top suction, single-stage pump with mixed flow type impeller is the selected design option for the secondary sodium pump (Figure 2.14) after detailed model tests in water. The pump is installed in a tank and has anticonvection

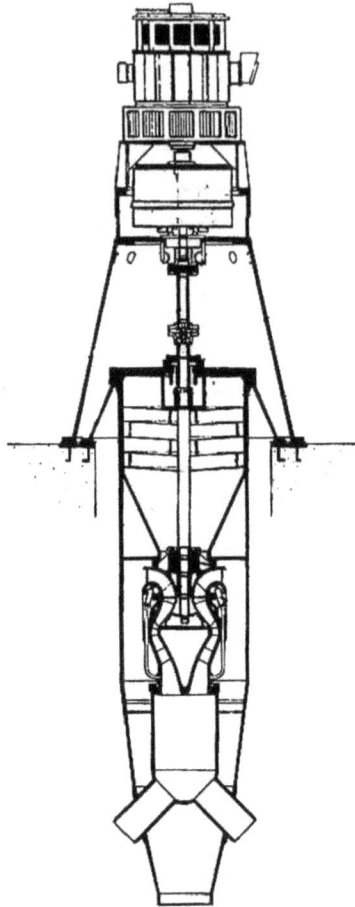

Figure 2.14 PFR secondary pump [12]. (Reproduced with kind permission of the International Atomic Energy Agency (IAEA)).

baffles in the cover gas space that minimise the heat transferred to the top flange keeping it at room temperature. An antivortex plate provided above the impeller suction prevents gas entrainment into the impeller.

2.3.5 Prototype Fast Breeder Reactor (PFBR) (India) [15, 16]

The Indian Prototype Fast Breeder Reactor (PFBR) uses two primary pumps in the primary circuit and one pump in each of the two secondary circuits. This choice of running the primary circuit with only two pumps for a large power reactor is unique and at variance with the arrangement in all other reactors where three or more pumps have been used. Furthermore, there is no check valve on the pump discharge line, and therefore in the eventuality of single pump operation, due to malfunction of the other unit, operation of the running pump is constrained to a narrow band of speed dictated from considerations of minimum bearing capacity and avoidance of cavitation.

2.3.5.1 Primary pump

The hydraulic design of the primary pump (Figure 2.15) ensures that there will be sufficient NPSH margin under the available suction head to avoid any cavitation erosion of the impeller over many years. The cavitation performance was extensively verified through water tests on 1/2.75 pump model, which allowed visual observation of the growth of cavitation bubbles. Furthermore, paint erosion method was used to study cavitation erosion in extended water tests to confirm cavitation erosion-free pump operation during service.

The impeller is of top suction type, and liquid sodium enters the impeller making a 180° turn after vertically rising through the space between the intake skirt and pump removable parts. An axial diffuser is employed to make the pump compact. High-pressure sodium leaving the diffuser enters the header connected to the core through the pump pipe connection. The shaft is of composite construction with a hollow middle portion connected at either end to solid forgings on which are mounted the journal of the hydrostatic bearing and the impeller at the bottom and the bearings and seals at the top. A six-pocket hydrostatic bearing radially supports the rotating assembly in sodium, while Kingsbury-type thrust bearing, and radial sleeve bearings support the rotating assembly outside sodium. The pump shaft is connected to the motor shaft using a disc type lamina coupling with ball transfer unit (flexible coupling). A disc-type flywheel is provided on the motor shaft to ensure a gradual reduction in pump flow rate (flow halving time of 8 s) in the event of loss of power supply. The total length of the shaft is ~11.3 m, and the total height of the pump (excluding motor assembly) is ~14.3 m. The total weight of the removable assembly is ~44 tons. Figure 2.15 is a sketch of the primary pump assembly. The pump is suspended from its top flange, bolted to the roof slab of the reactor. The bottom end of the pump engages with a conical receptacle, called pump pipe connection (PPC), connected to the piping leading to the core. The top flange of the pump is at 383K, while the bottom end is at 670K (cold pool sodium temperature) – this differential temperature results in the tilting of the pump during operation. The pump is therefore assembled at room temperature with a pre-determined inclination so that it is vertical at the rated operating temperature (670K). The inclination of the rotating assembly during part load operation of the reactor is achieved by spherical support (spherical plain bearing).

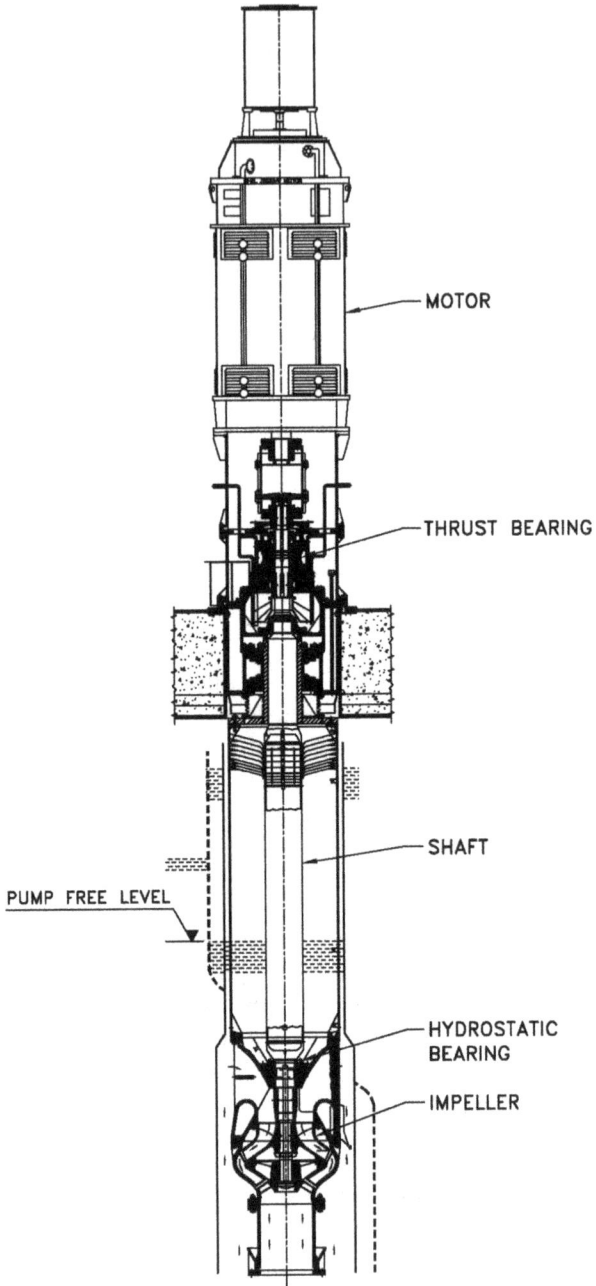

Figure 2.15 Primary pump of PFBR.

2.3.5.2 Secondary pump

The secondary sodium pump (Figure 2.16) of PFBR is a bottom suction, mixed flow pump with an axial diffuser. The pump is contained in a tank with a spherical bowl into which the pump discharges high-pressure sodium. The bowl has two symmetrical nozzles 180° apart and high-pressure sodium

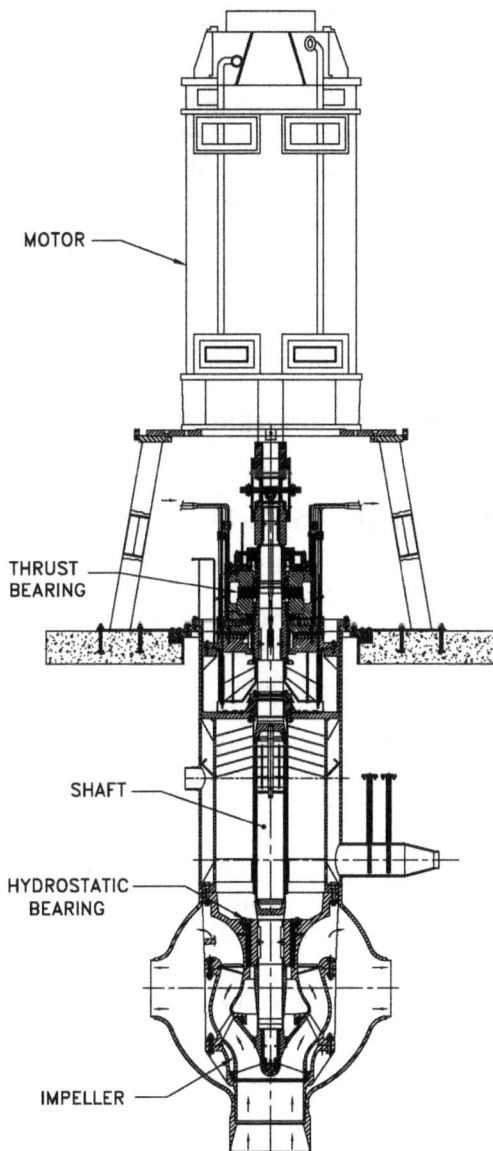

Figure 2.16 Secondary pump of PFBR.

exits through these nozzles and into the tube side of the two intermediate heat exchangers of the loop (two such independent loops constitute the secondary sodium system of PFBR.) The rotating assembly is supported in sodium by a six-pocket hydrostatic bearing supplied from the high-pressure spherical bowl. The high-pressure spherical bowl is separated from the pump tank's relatively lower pressure cover gas region by piston-ring type seals. The liquid leaking past these piston-ring type seals into the cover gas region overflows into the sodium storage tanks. An electromagnetic pump circulates the sodium through the purification circuit and feeds it back into the secondary system.

Oil-lubricated radial sleeve bearing and Kingsbury thrust bearing provide support at the top of the rotating assembly. Two mechanical seals are provided back-to-back between the pump cover gas and the Kingsbury bearing and an additional seal is provided to seal the bearing oil from the atmosphere. A gear type flexible coupling connects the pump and motor shafts. A disc-type flywheel mounted on the motor shaft ensures gradual flow coastdown (flow halving time of 4 s) in the event of loss of power supply.

2.4 SODIUM PUMPS OF COMMERCIAL SIZE FAST REACTORS

2.4.1 SuperPhénix-1 (aka SPX-1) (France) [1, 3, 17]

2.4.1.1 Primary pump

The general design scheme of the SuperPhénix-1 pumps is similar to that of the Phenix pumps. However, the following distinctions in the overall hydraulics of the primary sodium pumps (Figure 2.18) deserve mention:

(i) The sodium hydrostatic bearing is placed beneath the impeller.
(ii) The diameter of the lower labyrinth seal is chosen to reduce considerably the hydraulic axial thrust such that the direction of the net axial thrust remains downward under all operating conditions.
(iii) The check valve (non-return valve) at the pump discharge in the earlier pumps is removed, and a shutter made of steel ring (*obturator* in French parlance) is introduced in the impeller suction passage.

The weight of the rotating assembly and bearing load is reduced by making the central region of the shaft hollow with a diameter of 600 mm and thickness of 20 mm. The axial distance between the top and bottom bearings is 10 m. The hydrostatic bearing is of pocket-throttle type but without

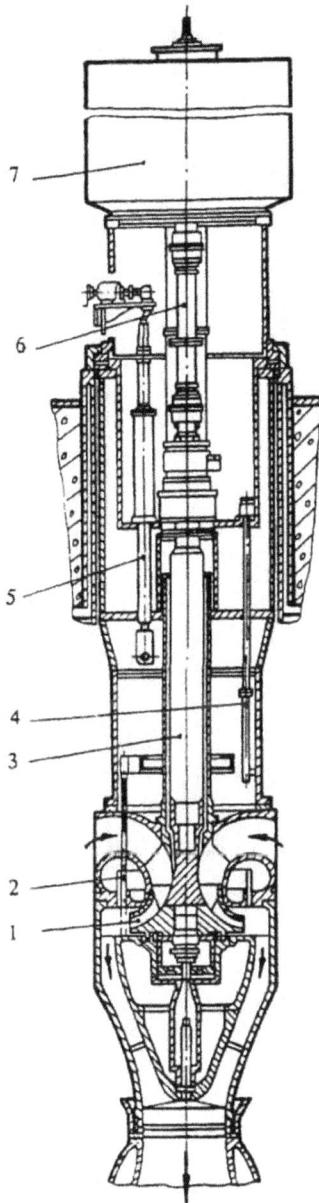

Figure 2.17 SuperPhénix-1 primary pump [3]. 1 – impeller; 2 – shutter; 3 – pump shaft; 4 – level gauge; 5 – shutter drive; 6 – connecting shaft; 7 – motor. (Reproduced with kind permission of the International Atomic Energy Agency (IAEA)).

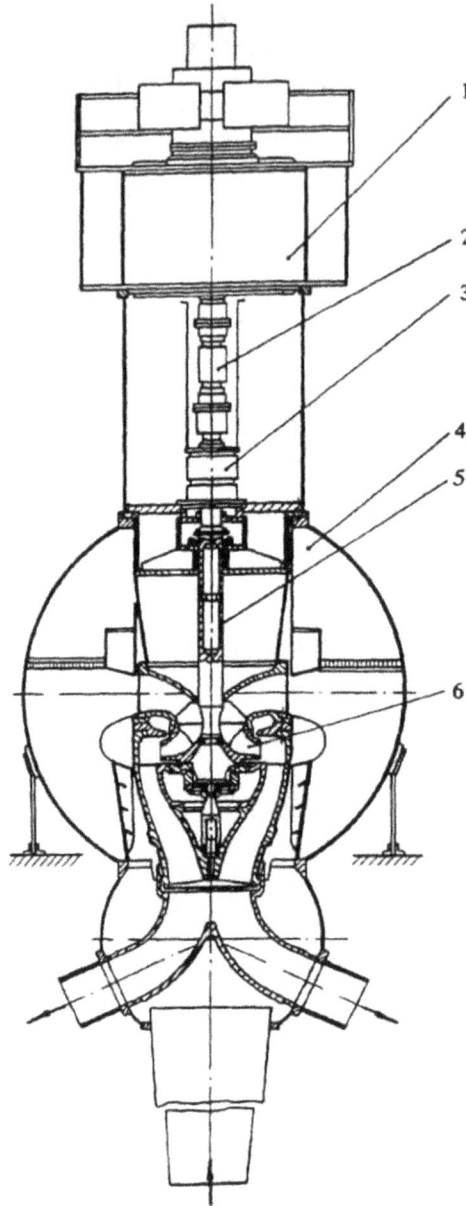

Figure 2.18 SuperPhénix-1 secondary pump [3]. 1 – motor; 2 – connecting shaft; 3 – shaft seal; 4 – expansion tank; 5 – pump shaft; 6 – impeller. (Reproduced with kind permission of the International Atomic Energy Agency (IAEA)).

discharge grooves between the pockets, unlike that in Phenix and other small pump bearings. The absence of grooves between the bearing pockets results in:

  (i) Equalisation of pressure between the pockets and increased hydrody-
      namic action.
 (ii) Reduction of the risk of jamming of journal and bush at lower bearing
      load capacity compared to the Phenix pump.
(iii) Ease of fabrication of the bearing journal.

The shaft seal and upper bearings design are similar to that in the Phenix pumps. Although the seal face diameters are increased, their surface rubbing speeds (linear) are retained at approx. 10 m/s because of the lower pump speed (maximum pump speed of 500 rpm). A unique design is used to accommodate the differential thermal expansion between the pump top support (located on the roof slab at room temperature) and the pump discharge pipe to grid plate pipe connection (which is at ~400°C).

2.4.1.2 Secondary pump

The secondary sodium pump (Figure 2.18) hydraulics is similar to that of the primary sodium pump, with the impeller and shaft being smaller than the primary pump (to suit the required ratings). However, the hydrostatic bearing is the same as in the primary pump. No biological shielding is present and thermal insulation is used to protect the top roller bearing and mechanical seal. The pump is installed in the expansion tank, which is located at the highest point in the loop. Figure 2.18 shows the secondary sodium pump.

2.4.2 SuperPhénix-2 (aka SPX-2) (France) [18]

2.4.2.1 Primary pump

The SuperPhénix-2 reactor envisaged using four primary pumps in a pool-type layout. The design of the pump hydraulics was optimised to reduce the size and weight of the pump by:

  (i) Reducing the margin on NPSHR to only 20%.
 (ii) Increasing the sodium flow velocity to 7 m/s.
(iii) Using a helico-centrifugal type impeller with more efficient impel-
      ler eye. The operating experience with the pumps of the Phenix and
      SuperPhénix-1 reactors helped make these modifications. The possibil-
      ity of reducing manufacturing costs by replacing castings (for some
      hydraulic parts) by fabricated parts was also considered.

The vertical primary pump (Figure 2.19) is a single-stage, top suction design with mixed flow impeller. The pump is rigidly located (suspended) on the reactor roof slab in contrast to the flexible ring support in SPX-1 to simplify the arrangement and reduce cost. Although this supporting arrangement is less favourable than that in SPX-1 for seismic loading, it is demonstrated that it is compatible with Operating Basis Earthquake (OBE) requirements for SPX-2. The high-pressure sodium exiting the pump is discharged to the reactor core diagrid plenum through an articulated sleeve (similar to that in

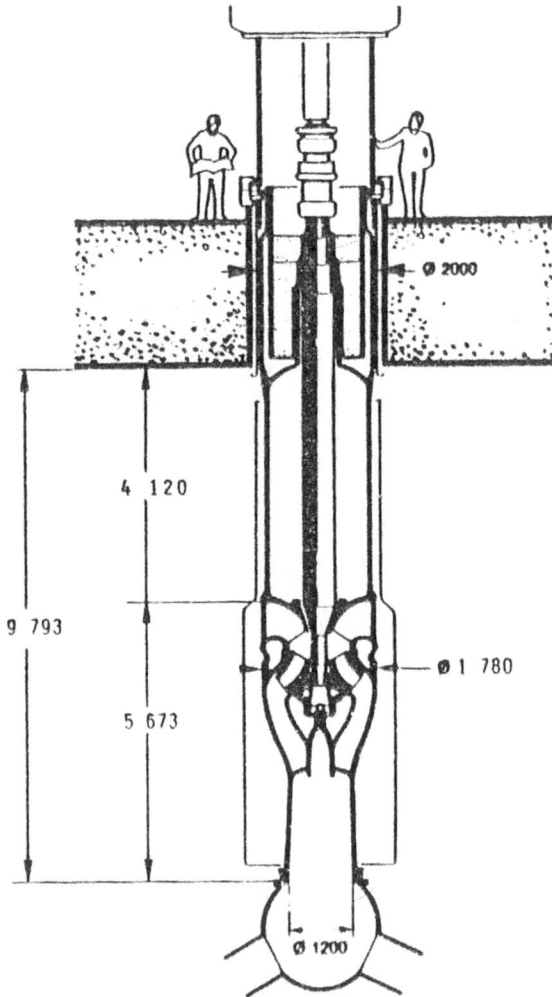

Figure 2.19 SuperPhénix-2 primary pump [18]. (Reproduced with kind permission of the International Atomic Energy Agency (IAEA)).

Phenix), hung on gimbals from the diffuser, and provided with seal rings at either end. This feature accommodates the differential expansion due to the difference in the roof slab and diagrid temperatures. A variable speed drive regulates the pump speed between 20% and 100% of the nominal speed. Unlike in the SuperPhénix-1 primary pump, no check valve or blocking skirt is provided to prevent reverse flow through the pump. However, an anti-reverse gear prevents turbining of the stopped pump under reverse flow. The flywheel is mounted on the motor drive shaft.

2.4.2.2 SPX-2 secondary pump

The secondary pump (Figure 2.20) is a mixed flow type with four impeller vanes and eight diffuse vanes. The high-pressure sodium from the diffuser is discharged into a spherical bowl with two symmetrical discharge nozzles, which are connected to the secondary piping.

The rotating assembly is supported in sodium by a hydrostatic bearing and at the top by oil-lubricated radial and thrust bearings. Mechanical seals are provided to maintain leak tightness of argon cover gas. These seals are interchangeable with those on the primary pumps.

2.4.3 BN-800 (Russia) [1, 2]

The BN-800 primary pump is similar to the BN-600 primary pump. It has a double suction impeller designed for a higher flow rate. The upper radial and thrust bearings of the BN-800 primary pump are identical to that of the BN-600 pump. Mechanical seals are used to maintain the leak tightness of pump cover gas. The design of the rotating seal ring of the mechanical seal of the BN-800 pumps is improved to minimise the risk of potential distortion of the working surface of the ring. Moreover, the sealing at the rotating ring face is achieved by using rubber gaskets of circular cross section instead of profiled sealing collars. This has enhanced the life of the seals to 30,000 hours. The friction pairs are graphite – graphite, and the seal is designed for a pressure of 2 MPa.

Secondary Pump: The secondary sodium pump (Figure 2.21) of the BN-800 reactor has a mixed flow impeller and axial diffuser. The high-pressure sodium exiting the axial diffuser enters the annular header between the pump tank and the removable internals, and from there to the delivery pipe. This arrangement simplified the design and made the pump lighter and more economical than the BN-600 secondary pump.

2.4.4 European Fast Reactor (EFR) pump (co-operative venture between France, Germany, Italy, Belgium, Netherlands, and UK) [19–21]

The European Fast Reactor was a collaborative project involving three countries that had active fast reactor programmes in the 1980s/1990s:

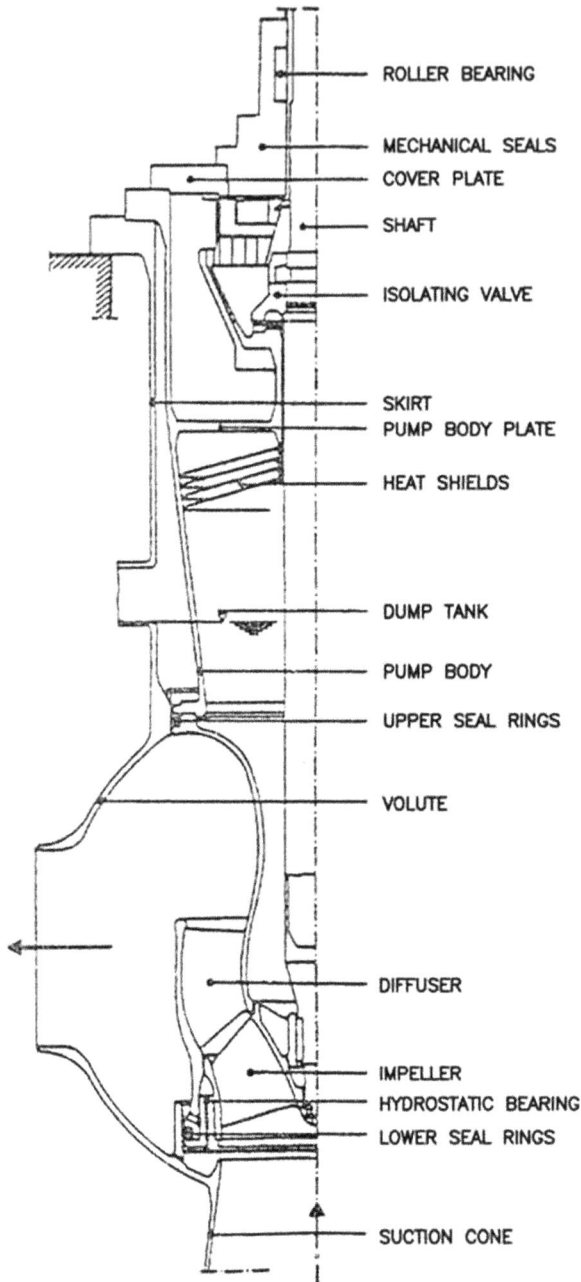

Figure 2.20 SuperPhénix-2 secondary pump [18]. (Reproduced with kind permission of the International Atomic Energy Agency (IAEA)).

Germany, UK, and France. Among them the French pump development was at the forefront as they had already constructed and operated the largest primary pump for SPX-1 reactor after building pumps for RAPSODIE and PHENIX reactors.

The EFR pump was largely based on the SPX-2 pump, which had been optimised to minimise the size and weight of the pump by adopting a fitness for-purpose cavitation criteria that permitted the use of a higher operating speed, compared to SPX-1 pump, despite having a larger capacity than it. The design envisaged three primary pumps and three secondary pumps.

While the German and UK pumps were simple upward flow bottom suction and double suction designs, respectively, the EFR design relied on a downward flow arrangement, i.e., single entry, top suction concept with a radial/axial diffuser. This arrangement permitted a high-pressure connection with a lower leakage area at the pump discharge nozzle. The disadvantages of the downward flow were increased eye diameter, as the inlet flow had to pass around the shaft and non-uniform velocity distribution at the impeller eye resulting from the 180° turning of the liquid entering the impeller.

However, as the requirement of absence of cavitation noise was forsaken, it was possible to sustain some decrease in the suction specific speed by reducing the NPSH margin towards the onset of cavitation. This permitted reduction in the overall dimension of the pump by increasing the rotating speed.

The EFR primary pump impeller was tested in IRIS loop (water), in EdF, and later in sodium in the CARUSO loop (one of the secondary loops of RAPSODIE). The NPSH values in water and sodium were found to be very similar. An erosion test at NPSH (d) + 10% margin (where NPSH (d) refers to the onset of head loss) of 1,360 hours showed slight traces of erosion on two blades of the impeller. The localised nature of the damage indicated that it depended upon the geometry of the blades, thus highlighting the importance of manufacturing tolerances on the onset of erosion. The cavitation criterion was then revised to the presence of 10 mm bubbles in a 1: 5 scale model. This relaxed limit on cavitation made it possible to keep the speed of the pump at ~530 rpm despite the capacity of the pump (29,500 m³/h) being larger than that of the Superphénix-2 primary pump (19,000 m³/h at ~ 700 rpm). This relaxation in the cavitation margin resulted in the maximum diameter of the EFR pump being less than that of the SPX-1 primary pump with a lower capacity of 16,000 m³/h.

The possibility of reducing manufacturing costs by using a flexible shaft, with the first critical speed in the operating range of the pump, was explored. Another innovative design option was also considered in light of the incident involving sodium contamination in the primary circuit of the Prototype Fast Reactor (PFR) due to oil leaking from the top bearing and seal. This option was the use of oilless bearings and seals (a combination of active magnetic bearing/gas bearing and ferrofluid seal).

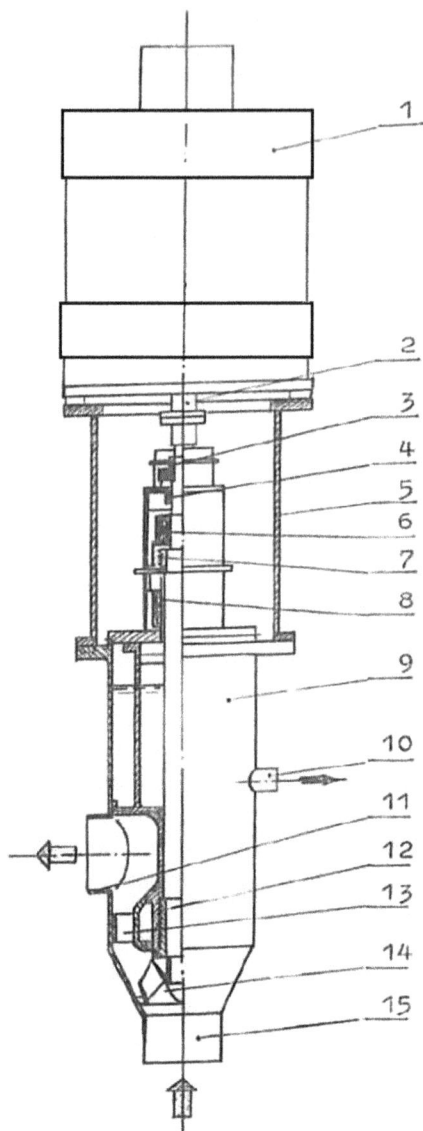

Figure 2.21 BN-800 Secondary Pump [2]. 1 – electric motor; 2 – clutch; 3 – axial bearing; 4 – radial bearing; 5 – electric motor mount; 6 – face seal; 7 – maintenance seal; 8 – shaft cooler; 9 – pump tank; 10 – sodium leakage outlet nozzle; 11 – delivery nozzle; 12 – hydrostatic bearing; 13 – guide device; 14 – impeller; 15 – suction nozzle. (Credit: Courtesy of the Nuclear Institute – Sketch taken from proceedings of Liquid Metal Engineering).

However, the SPX-2 and EFR pumps were never constructed because the fast reactor programme was suspended in the late 1990s in Europe, except Russia.

2.5 SUMMARY

This chapter familiarises the reader with the various types of centrifugal sodium pumps used in the different categories of fast reactors worldwide. There are several common traits in these pumps such as vertical construction, sodium hydrostatic bearing, shaft of partly hollow construction, argon cover gas over sodium free surface, etc. These and other special features will be discussed in more detail in the chapter on design.

NOTE

1 HNPF is a sodium-cooled, thermal reactor and NOT a fast reactor. These pumps were the earliest sodium centrifugal pumps successfully used and are therefore discussed here.

REFERENCES

1. F.M. Mitenkov, E.G. Novinsky and V.M. Budov, *Main Circulation Pumps for Atomic Power Stations*, edited by F.M. Mitenkov, member, Academy of Sciences, U.S.S.R., 2nd Edition, ENERGOATOMIZDAT, Moscow, 1990.
2. S.A. Belov, F.M. Mitenkov, E.G. Novinski, G.M. Nikolushkin and G.P. Shishkin, Design and experimental development of sodium pumps, Paper no. 121, Liquid Metal Engineering Technology, proceedings of the third international conference held in Oxford on 9–13 April 1984.
3. Status of Liquid Metal Cooled Fast Reactor Technology, IAEA-TECDOC-1083, April 1999.
4. B.P.S. Rao, S. Ghosh, M. Mannaru and R. Chandramohan, Manufacture of F.B.T.R. Sodium Pumps, Pumping Equipment in Nuclear Industry and Thermal Power Plants, *Proceedings of National Symposium*, ed. R.D. Kale, February 24–25, 1994, B.A.R.C., Bombay.
5. J.R. Davis, G.E. Deegan, J.D. Leman and W.H. Perry, Operating experience with sodium pumps at EBR-II, Report no. ANL/EBR-027, October 1970.
6. W. Babcock, State of Technology Study–Pumps: Experience with High Temperature Sodium Pumps in Nuclear Reactor Service and their Application to FFTF, Report no. BNWL-1049 UC-80, Reactor Technology, December 1969.
7. P.G. Smith, Experience with High Temperature Centrifugal Pumps in nuclear reactors and their application to molten salt thermal breeder reactors, ORNL-TM-1993, September 1967.
8. R.W. Atz and M.J. Tessier, High Temperature Testing of a Sodium Pump, Presented at *A.S.M.E. Winter Annual Meeting*, December 1978.

9. M.C. Zerinvary and J. Wasko, "F.F.T.F. Primary Pumps Point Way to Design for Tomorrow's Liquid Sodium Handling", *Power* (N.Y.), 120(7), 54–57.
10. M. Guer, W. Raczynski and G. Keyser, SNECMA-Bergeron Sodium Pumps Development Stage Seen in the Light of The Phenix Experiment, C109/74, Pumps for Nuclear Power Plants, University of Bath, 22–24 April 1974.
11. Fast Reactor Database 2006 Update, Report no. IAEA-TECDOC-1531, December 2006.
12. J.M. Laithwaite, L. Bowles and F.M. Delves, Sodium Pumps for Fast Reactors, Paper no. IAEA-SM-130/10, *Proceedings of a symposium on Progress in Sodium-Cooled Fast Reactor Engineering*, Monaco, 23–27 March 1970.
13. L.F. Bowles and D. Taylor, The Sodium Pumps for the P.F.R, Nuclear Engineering, May 1967.
14. J.M. Laithwite and A.F. Taylor, Hydraulic Problems in the P.F.R. Coolant Circuit, *Proceedings of a symposium on Progress in Sodium-Cooled Fast Reactor Engineering*, Monaco, March 1970.
15. S.G. Joshi, A.S. Pujari, R.D. Kale and B.K. Sreedhar, Cavitation Studies on a Model of Primary Sodium Pump, Proceedings of FEDSM'02, *The 2002 Joint US ASME European Fluids Engineering Summer Conference*, July 14–18, 2001, Montreal, Canada.
16. K.V. Sreedharan, S. Athmalingam, V. Balasubramaniyan, A.S.L.K. Rao, P. Chellapandi and S.C. Chetal, Design and Manufacture of Sodium pumps of 500 MWe Prototype Fast breeder Reactor (P.F.B.R.), *20th Annual Conference of Indian Nuclear Society*, January 4–6, 2010, Chennai, India.
17. H. Noel and G. Pasqualini, Fabrication and Testing of Main Sodium Pumps of Super Phénix-1, *Nuclear Technology*, 68: 2, 153–159, 1985. DOI: 10.13182/NT85-A33551.
18. Roger Morriset, Maurice Gravier, Jean-Marc Canini, Pool Type Breeder Primary and Secondary Sodium Pumps: Design Developments, *Proceedings of Symposium on Fast Breeder Reactors: Experience and Trends, Lyons*, 1985.
19. W. Marth, A Review of the Collaborative Program on the European Fast Reactor (EFR), Presented at *the Annual Meeting of IWGFR*, April 1992.
20. Unusual Occurrences During LMFR Operation, Proceedings of a *Technical Committee meeting held in Vienna*, 9–13 Nov. 1998, Report no. IAEA-TEC DOC-1180.
21. A.A. Rinejski, Fast Breeder Reactors: Advanced Concepts for Economic Design, SCUAE, USSR, April 1990.

Chapter 3

Design of centrifugal sodium pumps

3.1 INTRODUCTION

This chapter discusses the various factors that govern the hydraulic and mechanical design of pumps for sodium-cooled reactor applications. There are excellent textbooks/papers on the design of pumps [1–10], and a discussion on similar topics will be redundant and repetitive. Instead, we focus on the various alternatives available to the designer who embarks on the design exercise, possibly for the first time, and the thinking that goes into arriving at a final design configuration. Examples are provided wherever necessary to clarify the concept discussed.

The heat transport system of a fast reactor consists of a primary (radioactive) sodium circuit, an intermediate secondary (non-radioactive) sodium circuit, and a tertiary (conventional) steam water circuit. The primary heat transport system can be either a loop-type configuration or a pool-type configuration, as discussed in Chapter 1.

The conceptualisation of the basic design configuration is the first stage in the design process. The designer must first decide on the location of the pump (hot or cold leg). This decision will provide input for selecting suitable materials for the construction of the pump. The pump configuration (e.g., single/double suction impeller, top/bottom suction impeller, volute casing/diffuser pump) is based on the estimated flow rate and pressure drop in the system. The pump's operating parameters (i.e., pump flow rate, head, operating speed) are finalised based on the hydraulic and space limitations (e.g., NPSH available, single/double suction, restriction on maximum diameter). The detailed design that follows takes into consideration the mechanical features such as type of shaft (solid/hollow composite geometry), type of bearings and seals, coupling, etc., to achieve an economical and robust design with ease of maintenance. The detailed hydraulic and mechanical design then follows.

DOI: 10.1201/9781003460350-3

3.2 DESIGN OPTIONS

3.2.1 General

3.2.1.1 Pump location

The option of locating the pump in either the hot or cold leg is available for loop-type reactors in both the primary and the secondary circuits. In a pool-type reactor, however, the possibility exists in only the secondary system because the primary pump is invariably in the cold pool.

It should be noted that the pump, when placed in the cold leg, has several advantages, which include:

(a) easy material selection due to the pump operating temperature (400°C) being below the creep temperature of the material;
(b) simplified cooling arrangement of the top bearing and seals because of reduced heat transfer to these components; and
(c) low susceptibility to thermal transients because of the presence of IHX (in the case of the primary circuit) or the steam generator (in the case of the secondary circuit), upstream of the pump, that dampens thermal shock effect in the event of a transient. In this arrangement, however, the additional pressure drop in the IHX/steam generator lowers the NPSH available at the pump suction, thereby making it necessary to increase the pump submergence resulting in a longer shaft span between bearings.

On the contrary, the hot leg pump necessarily operates above the creep temperature because the operating temperature depends on the reactor outlet temperature (530°C–550°C). Moreover, the cold thermal transients seen by the pump in case of reactor trip can be as high as 150°C in 10–15 seconds, i.e., at 10°C to 15°C per second. Such thermal shocks are critical for parts operating with low clearances, such as hydrostatic bearings and wearing rings. However, the distinct advantage of locating the pump in the hot leg is a significant reduction in the pump suction side head losses because the IHX/steam generator is now downstream of the pump. This advantage permits lower submergence of the impeller/lower cover gas pressure and shorter shaft span between bearings compared to the cold leg pump. The net effect, therefore, of placing the pump in the 'hot' leg is to reduce the overall pump height. For example, in the BOR-60 primary pump, the shaft length could have been reduced to about 1.2 m, against the actual length of 5 m, had the pump been located in the hot leg instead of in the cold leg. In the final analysis, the overwhelming advantages resulting from locating the pump in the cold leg has favoured this option in the majority of reactors. Most reactor pumps have chosen the cold leg location mainly due to the ease of thermal design. The exceptions, however, are the primary pumps in the KNK-II, FFTF, CRBRP, SNR-300, and SNR-2 reactors; the secondary pumps in all the reactors, except RAPSODIE, are located in the cold leg. Table A1.3 in Appendix 1 summarises the locations of pumps in various experimental, prototype and commercial reactors.

3.2.1.2 Material of construction

The material of construction of sodium pumps is austenitic stainless steel (AISI SS304L (CF3 grade casting) or AISI 316L (CF3M grade casting)/AISI 316LN (CF3M grade casting with nitrogen), which are preferred for their good ductility, high temperature strength, and corrosion resistance. The 304L variety is used for pumps operating below the creep temperature of 425°C (cold leg pumps) while the 316L/316LN varieties are used for pumps operating above 425°C (hot leg pumps).

3.2.2 Rating

3.2.2.1 Pump flow rate

The pump flow depends on the number of pumps operating in parallel and the location of the pump; i.e., hot leg or cold leg (for a loop-type/piped system). The rationale for deciding the number of loops is based on factors including space availability, reactor capital cost, operating costs, availability of technology for pumps of the desired rating consistent with the expected life of the pump. Two/three or even four parallel loops are normally provided from considerations of safety (redundancy). The parallel operation, however, places stringent restrictions on pump hydraulics such as identical and drooping H-Q characteristics with a minimum specified ratio of shut-off head to the operating head.

3.2.2.2 Pump head

The pump head depends on the location of the pump; i.e., whether it is in the hot or cold leg for a piped system/loop-type configuration and the system head loss, which in turn depends on the system configuration.

3.2.2.3 Operating speed

The pump operating speed is selected on the basis of the following considerations:

(a) The maximum pump lateral dimension and, therefore, the diameter of the main vessel for a pool-type reactor or the diameter of the pump tank in the piped arrangement is inversely proportional to the pump speed for a given pump head.
(b) Suction specific speed: The suction specific speed of a pump is the capacity of the pump to operate under adverse NPSH conditions and is given by the relation,

$$\text{Nss}_{plant} = N\sqrt{Q}/\text{NPSHA}^{3/4} \text{ and } N_{ss\,reqd} = N\sqrt{Q}/\text{NPSHR}_{3\%}^{3/4}$$

where Nss_{plant} = plant/operating suction specific speed, $N_{ss\ reqd}$ = required suction specific speed, Q is pump flow rate in m^3/s, $NPSHR_{3\%}$ = Net Positive Suction Head Required at 3% head drop, NPSHA = Net Positive Suction Head Available, N is pump speed in rpm. The decision to locate the pump in the hot/cold leg decides the upper bound on the plant NPSH. An increase in plant NPSH is achieved either by increasing the submergence (which is not preferred because it increases the span between bearings and lowers the critical speed) or increasing the cover gas pressure. High cover gas pressure makes the sealing of cover gas difficult and is therefore not used, especially in the case of pumps in the primary (radioactive) loop. Based on the margin on plant NPSH and a realisable value of the required suction specific speed (~ 200 – with Q, N and $NPSHR_{3\%}$ in the units given above – is a reasonable value), the operating speed is worked out.

(c) Specific speed: Is defined as the speed of a geometrically similar pump which delivers unit quantity of liquid at unit head. It is given by the relation $Ns = N\sqrt{Q}/H^{3/4}$ where Ns = specific speed, Q is pump flow rate in m^3/s, H is pump head in m, N is pump speed in rpm. The overall pump efficiency increases with increase in pump specific speed primarily due to the reduction in friction losses and volumetric losses in the pump. The pump operating speed is selected so that the operating specific speed is in the range where the efficiency is the highest.

3.2.3 Hydraulics design

3.2.3.1 Type of impeller

The impeller can be of either single suction type (Figure 3.1) or double suction type (Figure 3.2). In a single suction impeller, all the flow rate enters one side of the impeller. In contrast, in a double suction impeller, the flow rate is divided into two streams that enter the impeller from either end. Double suction impellers have the following advantages:

(a) the flow rate entering each side of the impeller is half the total flow rate handled by a single suction impeller, and so the NPSH requirement of the double suction impeller is about 63% of that of a single suction impeller of the same suction specific speed; and

(b) the hydraulic axial thrust on the rotating assembly is almost zero. Double suction (Figure 3.2) impellers are used in the primary pumps of the Clinch River Breeder Reactor Project (CRBRP), in the primary and secondary pumps of Sodium Reactor Experiment Power Plant (SRE-PP), and the primary pumps of PFR, BN-600 and BN-800 reactors.

3.2.3.2 Top suction or bottom suction

In a vertical pump with single suction impeller, two arrangements are possible: bottom suction (Figure 3.1) or top suction impeller layout (Figure 3.3). The bottom suction arrangement has the advantage of increased submergence

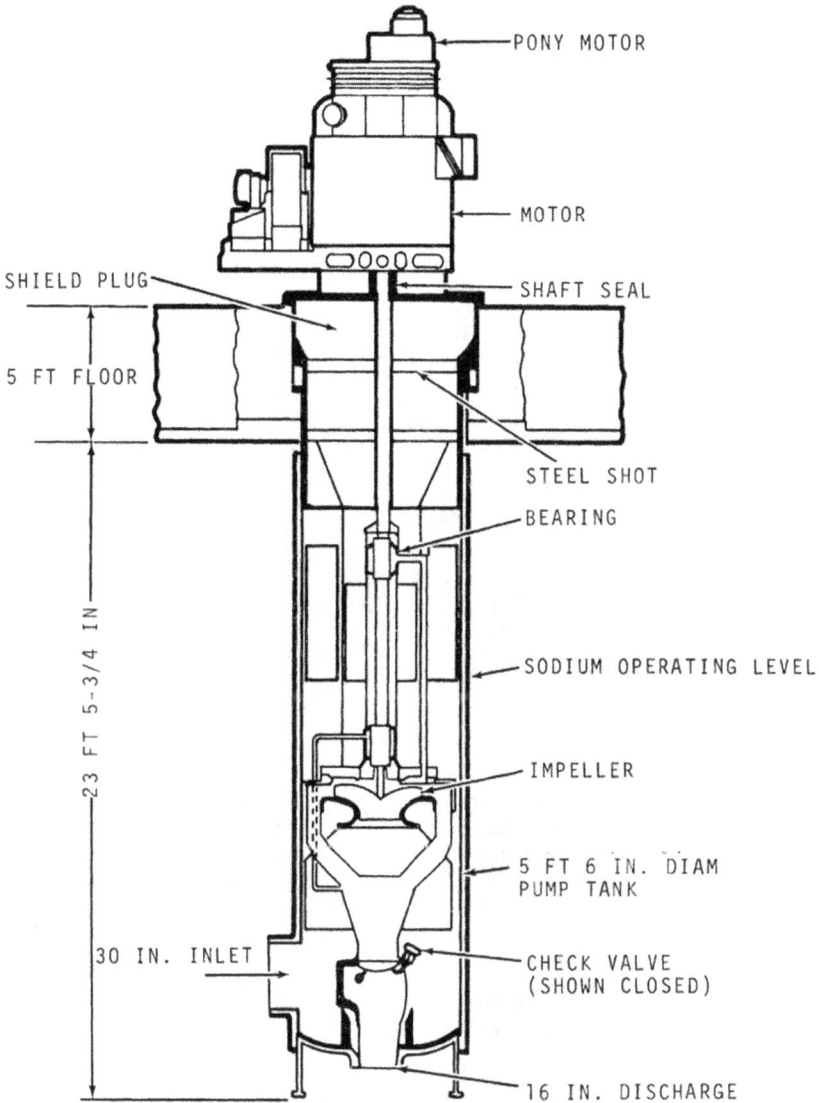

Figure 3.1 EFAPP primary sodium pump (single suction, bottom suction impeller) [11]. (Reproduced from W. Babcock, State of Technology Study – Pumps: Experience with high temperature sodium pumps in nuclear reactor service and their application to FFTF, AEC Research and Development Report no. BNWL-1049 UC-80, Reactor Technology, Dec. 1969 BNWL – Batelle Northwest Laboratories, Richland, Washington, USA).

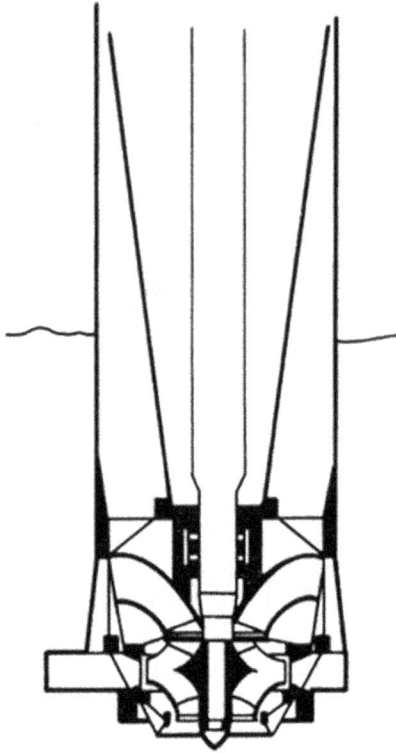

Figure 3.2 PFR primary sodium pump (double suction impeller) [11]. (Reproduced from W. Babcock, State of Technology Study – Pumps: Experience with high temperature sodium pumps in nuclear reactor service and their application to FFTF, AEC Research and Development Report no. BNWL-1049 UC-80, Reactor Technology, Dec. 1969BNWL – Batelle Northwest Laboratories, Richland, Washington, USA).

(NPSH available); however, the arrangement makes the discharge layout complicated especially for pool-type reactor pumps. On the contrary, while the top suction arrangement simplifies the discharge layout, it comes at the expense of the available NPSH, which is reduced due to the marginal reduction in impeller submergence and the additional losses incurred from the 180-degree turn of the liquid into the impeller eye. The designer has to judiciously select the final configuration based on confidence level in hydraulic design, experience with similar designs, etc. Examples of bottom suction design are primary pumps of BOR-60, EBR-II, FFTF, EFAPP, HNPF, JOYO, KNK, SNR-300; secondary pumps of PFBR, EFAPP etc. Examples of top suction impellers are those in the primary pumps of Rapsodie, Phenix, SPX, PFBR, the primary and secondary pumps of FBTR, secondary pumps of PFR etc.

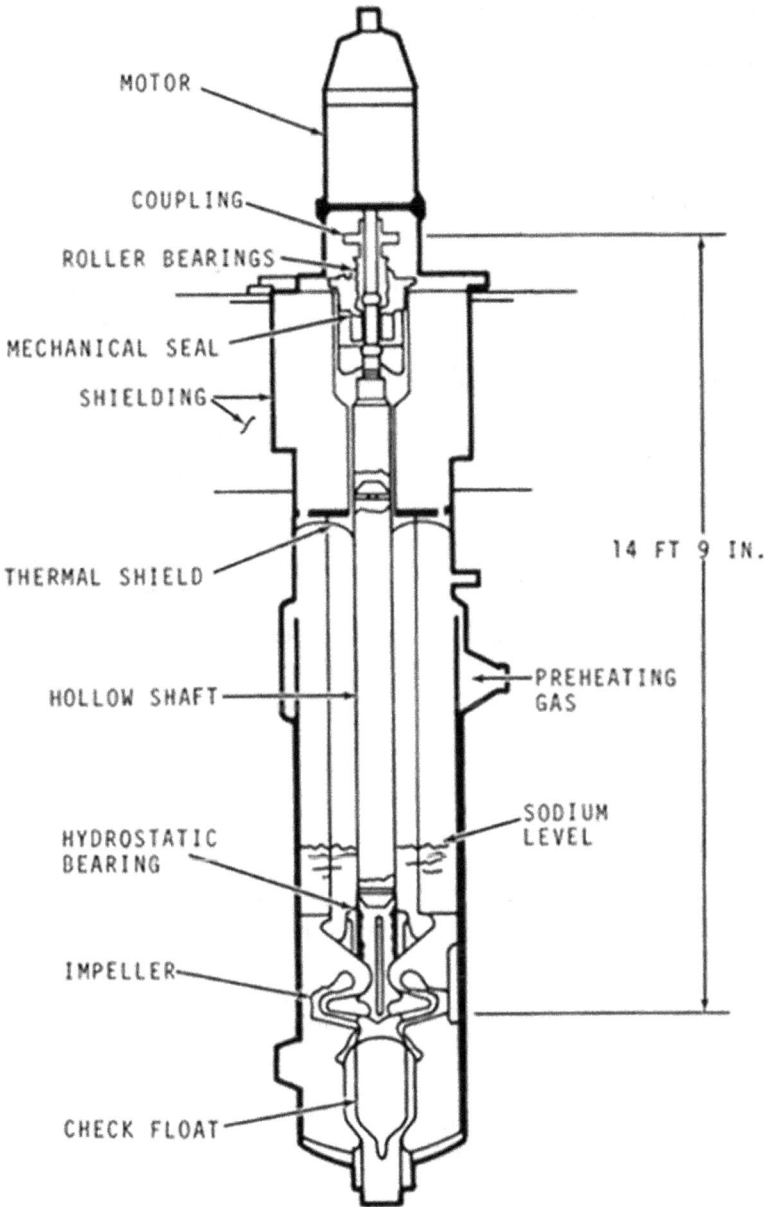

Figure 3.3 RAPSODIE primary sodium pump (single suction, top suction impeller) [11]. (Reproduced from W. Babcock, State of Technology Study – Pumps: Experience with high temperature sodium pumps in nuclear reactor service and their application to FFTF, AEC Research and Development Report no. BNWL-1049 UC-80, Reactor Technology, Dec. 1969 BNWL – Batelle Northwest Laboratories, Richland, Washington, USA).

3.2.3.3 Piped suction or sump suction

In the case of pumps for loop-type systems, there are two possible hydraulic configurations:

1. With the pump suction connected to a sump or pump tank.
2. With the pump suction connected to the piping system and the pump delivery to the pump tank (bowl) and the discharge piping.

In configuration 1, the piping is connected to the sump (or pump tank), and the NPSHA is a function of the sump/pump tank cover gas pressure and the impeller submergence. The losses in the upstream piping and IHX are, therefore, not charged against the NPSHA. However, the pump shaft length is governed by the losses between the reactor vessel and the sump (or pump tank). The shaft length is, therefore, controlled by this pressure drop rather than the NPSHA. With configuration 2, special seals are to be provided to separate the pump delivery, which discharges to the pump bowl/tank, from the lower pressure cover gas region downstream. This configuration is limited to a single suction impeller; therefore, the NPSHA controls pump shaft length. No such restriction exists for the sump-type arrangement; thus, as flow rates increase, moderate shaft lengths can be achieved (in the sump arrangement) with double suction impellers, requiring considerably less NPSH.

The primary sodium pumps of EFAPP, RAPSODIE, and FBTR are examples of configuration 1, while the primary sodium pump of HNPF and the secondary sodium pump of PFBR are examples of configuration 2.

3.2.3.4 Multi-staging

Just as parallel operation is an option to reduce the flow rate handled by the impeller, multi-staging of impellers reduces the head per stage/impeller, thus reducing the loading on the blades. This arrangement, however, complicates the system and is not popular for reactor use. Examples of multi-stage sodium centrifugal pumps are the two-stage primary pump and four-stage secondary pump used in the 2000 kW sodium pump test facility at the Los Alamos Scientific Laboratory (LASL) – Figure 3.4 [11]. Another notable exception is the primary pump of the Commercial Demonstration fast Reactor (CDFR) in UNK which is a two stage design.

Tables 3.1 and 3.2 give the pump flow rate, head, speed, and specific speed of primary and secondary coolant pumps in some important experimental, prototype, and commercial size fast reactors [12].

3.2.3.5 Bladed diffuser or volute casing

The conversion of velocity head generated by the impeller to pressure head occurs in the volute casing/diffuser. A volute casing offers the advantage of

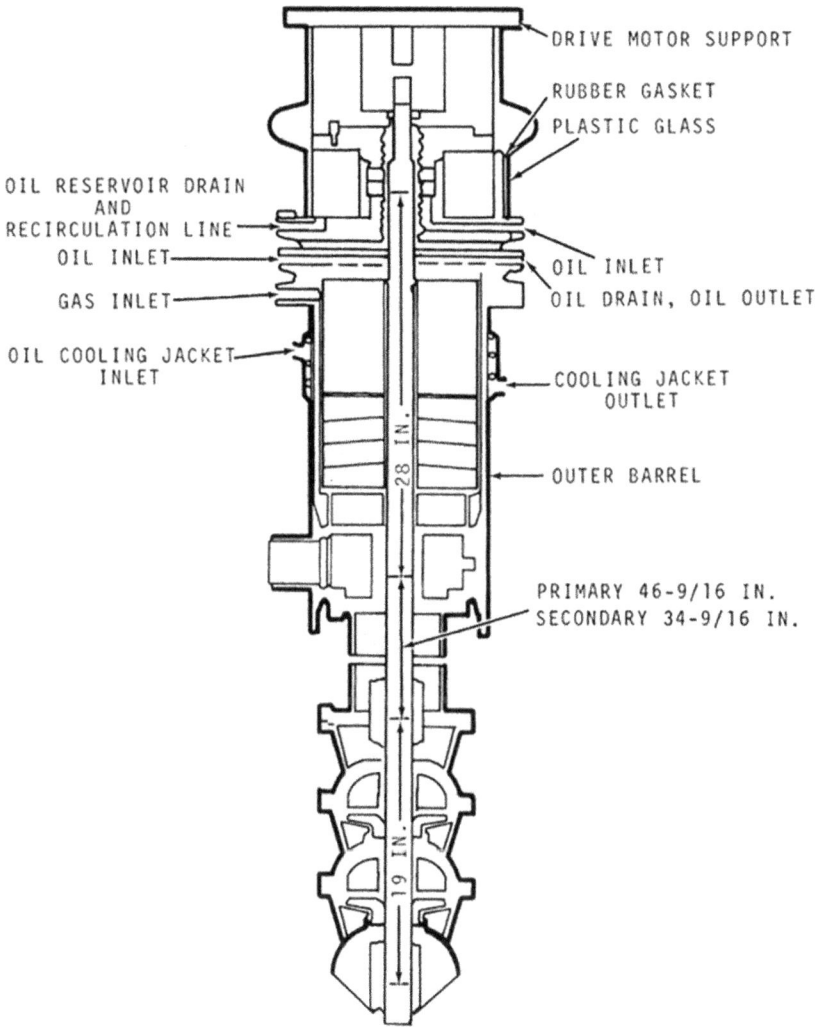

DRIVE MOTOR SUPPORT

RUBBER GASKET

PLASTIC GLASS

OIL RESERVOIR DRAIN
AND
RECIRCULATION LINE

OIL INLET

GAS INLET

OIL INLET

OIL DRAIN, OIL OUTLET

OIL COOLING JACKET
INLET

COOLING JACKET
OUTLET

OUTER BARREL

28 IN.

PRIMARY 46-9/16 IN.
SECONDARY 34-9/16 IN.

19 IN.

Figure 3.4 Multi-stage sodium pump in test facility at LASL [11]. (Reproduced
from W. Babcock, State of Technology Study – Pumps: Experience
with high temperature sodium pumps in nuclear reactor service and
their application to FFTF, AEC Research and Development Report no.
BNWL-1049 UC-80, Reactor Technology, Dec. 1969BNWL – Batelle
Northwest Laboratories, Richland, Washington, USA).

simplicity; however, in a volute casing, a net radial thrust is produced as the
liquid is slowed down. On the contrary, with a bladed diffuser, the radial
thrust is almost eliminated, and it is therefore preferred. Diffuser pumps
are, therefore, widely used (e.g., Rapsodie pumps, FBTR pumps, PFBR
pumps). Example of a volute-type pump casing is in the secondary pump of

Table 3.1 Pump rating – experimental fast reactors [12]

	Pump capacity (m³/min) Head (MPa) Speed (rpm) Specific speed (N in rpm, Q in m³/s, H in m) Motor power (kW)	
	Primary	Secondary
Rapsodie (France)	10.2	9.4
	0.46	0.25
	1250	1000
	25.6	30.6
	120	54
KNK-II (Germany)	10	8.6
	–	–
	1430	1430
	–	–
	–	–
FBTR (India)	11.0	6.2
	0.46	0.3
	1300	1450
	27.7	32.7
	150	55
JOYO (Japan)	26x2[a]	23x2[b]
	0.51[c]	0.35[d]
	930	1060[e]
	28.5	40.9
	330	220
BOR-60 (Russian Federation)	10	~14.0
	0.85	0.6
	1200	1200
	15.6	24.5
	285	–
EBR-II (USA)	34.1	22.3
	0.386	–
	880	–
	37.8	–
	260	–
EFAPP (USA)	45	49
	1.03	0.40
	875	900
	21	46.2
	1000	350

(*Continued*)

Table 3.1 (Continued) Pump rating – experimental fast reactors [12]

FFTF (USA)	56	56
	1.01	0.81
	1100	1110
	28.7	35.5
	1520	1110
CEFR (China)	14.25	9.5
	0.38	0.35
	900	900
	25.4	22.3
	150	150

[a] 21×2 in MK-I, II
[b] 21×2 in MK-I, II
[c] 0.63 in MK-I, II
[d] 0.37 in MK-I, II
[e] 975 in MK-I, II

Table 3.2 Pump rating – demonstration or prototype/commercial fast reactors [12]

	Demonstration reactors	
	Pump capacity (m³/min) Head (MPa) Speed (rpm) Specific speed (N in rpm, Q in m³/s, H in m) Motor power (kW)	
	Primary	Secondary
Phenix (France)	63	52
	0.5	0.4
	820	800
	39.3	41.6
	800	500
PFBR (India)	258	198
	0.61	0.55
	590	900
	48.1	71.6
	3600	2600
MONJU (Japan)	100	71
	0.8	0.5
	850	1100
	36	56.7
	2000	800

(Continued)

Table 3.2 (Continued) Pump rating – demonstration or prototype/commercial fast reactors [12]

	Demonstration reactors	
	Pump capacity (m³/min) Head (MPa) Speed (rpm) Specific speed (N in rpm, Q in m³/s, H in m) Motor power (kW)	
	Primary	*Secondary*
PFR(UK)	84	75
	0.8	0.4
	950	950
	36.9	58.9
	4920	2010
CRBRP (USA)	130	115
	1.12	0.86
	1170	963
	42.7	41.9
	3940	3940
BN-350 (Kazakhstan)	53.3	63.3
	0.94	0.58
	1000	1000
	28.1	44.1
	1700	1100
BN-600 (Russian Federation)	161.71	133.3
	0.81	0.31
	1000	1000
	53.8	101.3
	3150	1330
	Commercial reactors	
SuperPhénix-1 (France)	290	230
	0.53	0.25
	433	470
	42.6	73.1
	4170	1620
DFBR (Japan)	191	156
	0.8	0.48
	855	875
	50.1	68.8
	3400	900

(Continued)

Table 3.2 (Continued) Pump rating – demonstration or prototype/commercial fast reactors [12]

BN-800(Russian Federation)	205	192
	0.82	0.42
	990	990
	59.5	96
	4300	2000
EFR	450	177
	0.6	0.457
	530	780
	59.2	67.7
	To be determined	1660

the EFAPP reactor (Figure 3.5) as well as the primary and secondary sodium pumps of the BN-600 reactor (Figures 2.10 and 2.11).

3.2.3.6 Non-return valve (NRV)

Non-return valve (NRV) at the outlet of the primary pump is provided in the case of pumps operating in parallel to prevent flow from bypassing the core and passing through the stopped pump/pumps in the event of operation with one or more pumps not operational. In such cases, using a non-return valve increases the operating flexibility by making the operation of pumps possible over a wide speed range. However, the submergence of the impeller is reduced to the extent of space used to accommodate the valve. NRV is provided in primary pumps of RAPSODIE (Figure 3.3), FBTR, Phenix, BN-600 reactors. As given in Sec 2.4, in the SPX-1 reactor a shutter (aka obturator in French parlance) is provided at the pump suction instead of a check valve at the pump discharge.

In the PFR, a valve is provided in the pump discharge pipe that functions as a combined non-return and shut-off type valve. In the event of a pump trip, the pressure at the discharge of the operating pumps will hold the poppet against the seat and prevent flow through the tripped pump. A dashpot is integrated with the valve to avoid sodium hammer. The valve also functions as a shut-off valve when it is closed during power operation by manually operating the central spindle above the reactor's roof slab [13].

3.2.4 Mechanical design

3.2.4.1 Bearings in sodium

Conventional antifriction bearings are not used in sodium because of the poor lubricating properties of liquid sodium. A hydrostatic bearing, which uses the pressurised sodium from the pump outlet to provide bearing action, is, therefore, the bearing of choice.

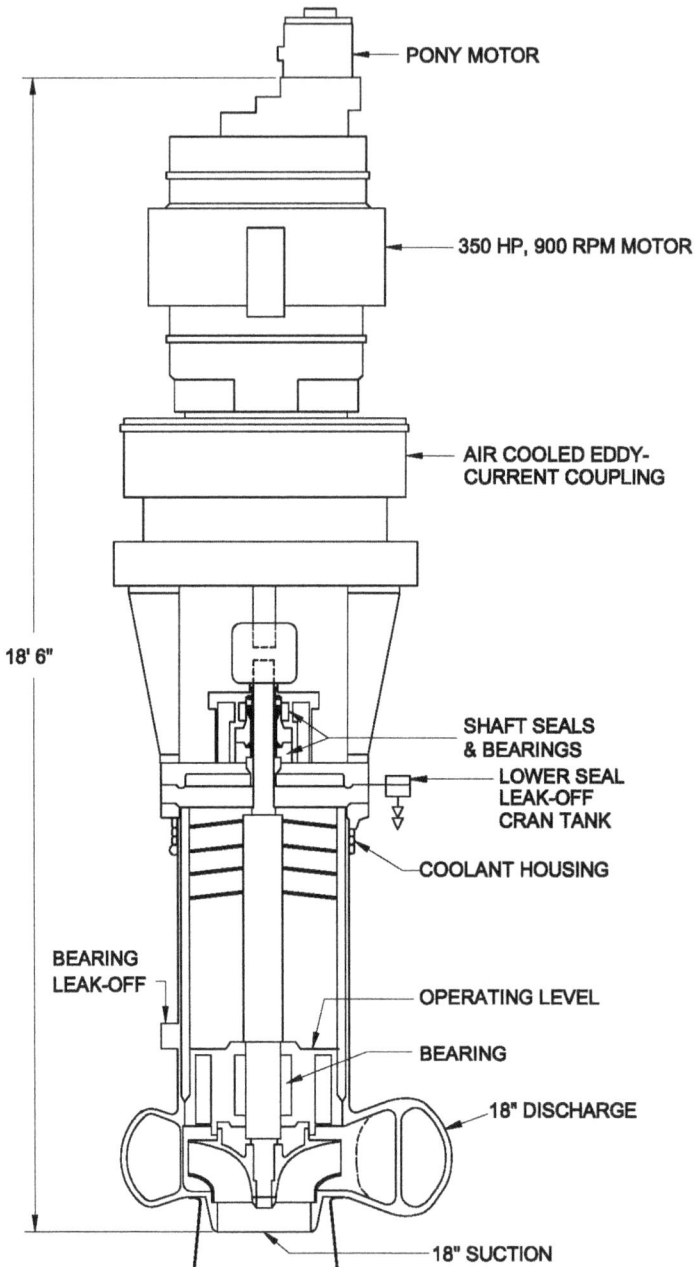

Figure 3.5 Secondary pump of EFAPP reactor. This is a volute-type pump [11].
(Reproduced from W. Babcock, State of Technology Study – Pumps:
Experience with high temperature sodium pumps in nuclear reactor
service and their application to FFTF, AEC Research and Development
Report no. BNWL-1049 UC-80, Reactor Technology, Dec. 1969BNWL –
Batelle Northwest Laboratories, Richland, Washington, USA).

3.2.4.2 Bearings outside sodium

The rotor is supported outside sodium using conventional antifriction bearings (angular contact ball bearings, tapered roller bearings, tilting pad type bearings, etc.).

3.2.4.3 Shaft of composite geometry

In a pool-type reactor, the pump lateral dimension directly influences the pool size (and consequently the reactor capital cost). It is, therefore, desirable to design the pump for the maximum speed achievable from cavitation consideration. The NPSH available is maximised by providing as high submergence as possible. However, this results in a large span between the bearings resulting in a long shaft. The shaft is designed to ensure that the critical speed of the rotating assembly is higher than the operating speed by ~25%. The required margin on the critical speed, above the operating speed, can be achieved while keeping the weight minimum by using a shaft of composite geometry, viz., a shaft that is solid at the impeller and bearing regions (bottom and top) and hollow in the central portion. Sodium pumps built so far have used only 'rigid rotor' wherein the pump operating speed is below the first critical speed. Interestingly, deviating from conventional thinking, a flexible shaft was considered in the design of the primary pump of the SPX-2 reactor to reduce the weight and cost of the pump. This project, however, did not progress beyond the drawing board due to the abrupt stoppage of the fast reactor programme in France.

3.2.4.4 Rotor support configuration

The following support arrangements have been used:

(a) Four bearings configuration with flexible coupling: This arrangement consists of two bearings in the pump and two in the motor. In the pump, the bearing in sodium is a radial, hydrostatic bearing, and the one at the top of the rotor assembly (in the shaft seal region) is oil-lubricated thrust and radial bearing. The motor shaft is provided with an oil-lubricated thrust bearing at the top and a similarly lubricated radial bearing at the bottom. The shaft and motor are coupled using a flexible coupling.

   This arrangement is most common and used in, for example, the pumps in HNPF (Figure 3.6), FFTF as well as the PFBR pumps.

(b) Four bearings configuration with quasi-flexible (e.g., diaphragm type) coupling: In this arrangement, the pump rotor assembly is supported in sodium by two hydrostatic bearings while the motor shaft is provided with oil-lubricated thrust bearing and radial bearing. No bearing is provided in the pump shaft seal assembly region. The CRBRP pump is

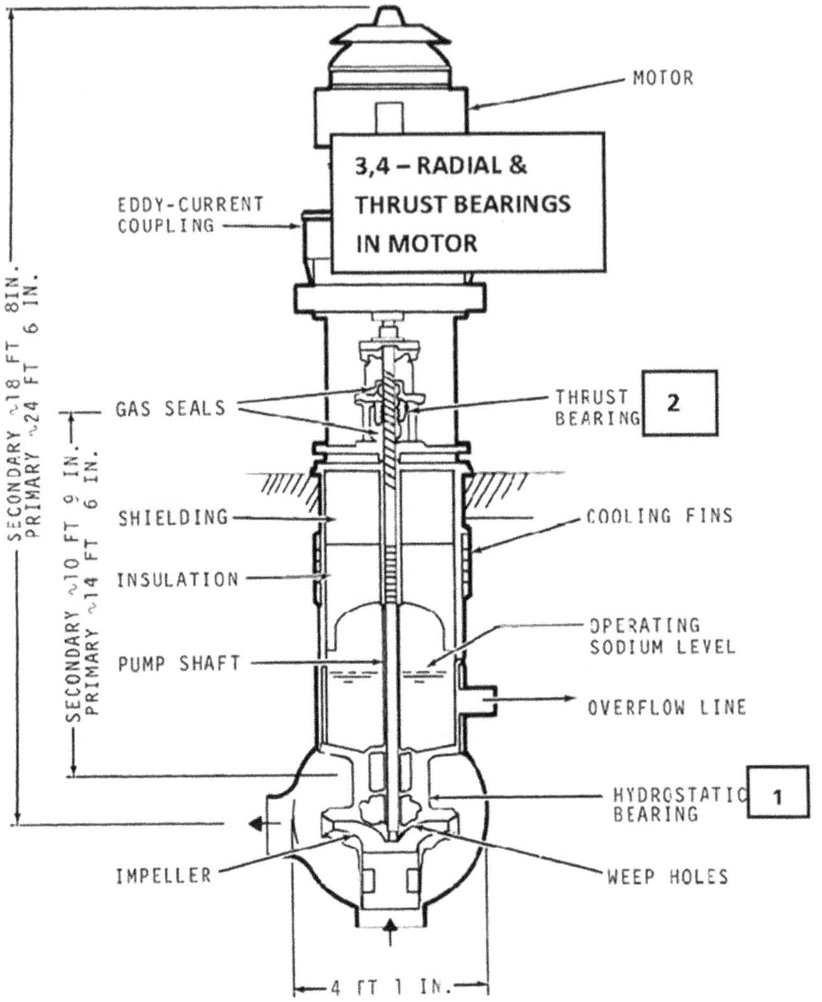

Figure 3.6 HNPF primary pump (example of bottom suction impeller. Also note the 4 bearings arrangement to support the pump rotor assembly) [11]. (Reproduced from W. Babcock, State of Technology Study – Pumps: Experience with high temperature sodium pumps in nuclear reactor service and their application to FFTF, AEC Research and Development Report no. BNWL-1049 UC-80, Reactor Technology, Dec. 1969 BNWL – Batelle Northwest Laboratories, Richland, Washington, USA).

an example of this arrangement. The rotor assembly of the EFAPP pump (Figure 3.1) also is supported similarly on two hydrostatic bearings in sodium.[1]

(c) Three bearings configuration: In this arrangement, the motor shaft is supported at the top by oil-lubricated radial and thrust bearings and

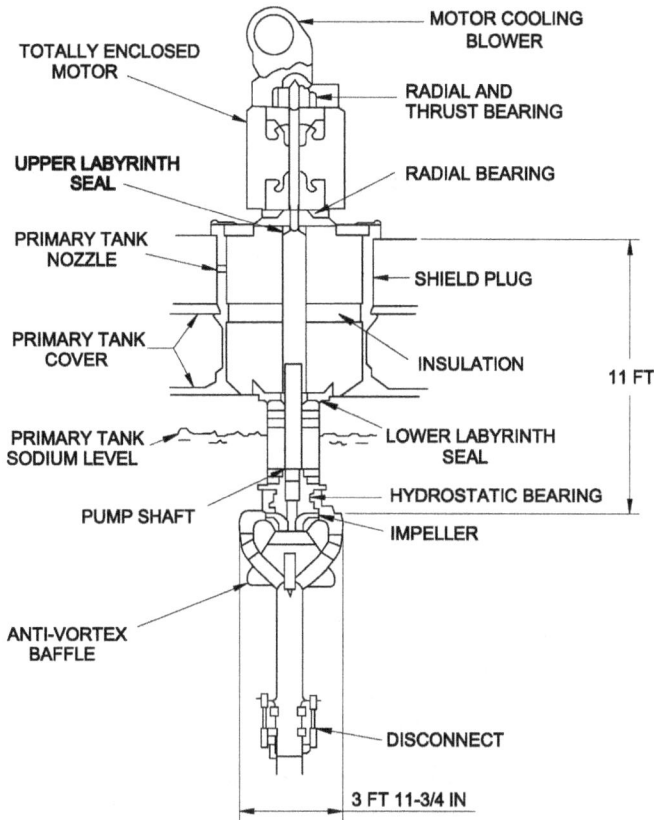

Figure 3.7 EBR-II primary sodium pump – three bearings arrangement [11]. (Reproduced from W. Babcock, State of Technology Study – Pumps: Experience with high temperature sodium pumps in nuclear reactor service and their application to FFTF, AEC Research and Development Report no. BNWL-1049 UC-80, Reactor Technology, Dec. 1969 BNWL – Batelle Northwest Laboratories, Richland, Washington, USA).

at the bottom by a radial bearing, while the pump rotor assembly is supported in sodium by a radial hydrostatic bearing. No bearing is provided in the pump shaft seal assembly region. The motor and pump shafts are connected with a rigid coupling. The EBR-II pump (Figure 3.7) is an example of this arrangement.

3.2.5 Pump support

In pool-type reactors, the primary pumps are suspended from thrust bearings fixed to the roof slab, while the discharge pipes at the bottom of the pumps are connected to the reactor diagrid pipes that lead the high-pressure sodium to the reactor core. The temperature of the cold pool where the

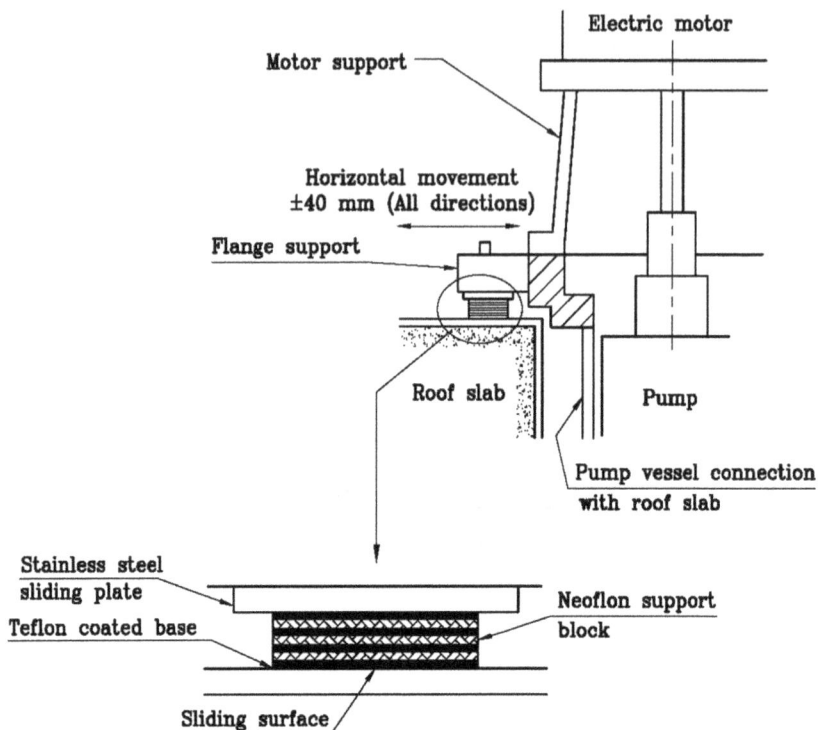

Figure 3.8 Arrangement to accommodate differential axial and radial thermal expansion in primary pump of Phenix [14]. (Reproduced with kind permission of the International Atomic Energy Agency (IAEA)).

primary pumps are located varies from 200°C to 400°C (depending on the reactor power) whereas the hot pool in which the core is located is at 550°C. Moreover the temperature of the roof slab of the reactor is ~60°C–80°C. As a result, the primary pump is subjected to a temperature gradient resulting in a differential expansion in both the radial and axial directions.

The pump top support, the pump flange to roof slab joint, and the connection of the pump discharge to the grid pipe are, therefore, designed to permit free thermal expansion, easy removal/insertion of the pump into the sodium pool (maintenance friendly), and vibration-free operation. The following paragraphs discuss the arrangements in various pumps.

3.2.5.1 Pump top support

In the Phenix primary pump [14], the assembly of the pump and motor is mounted on a base coated with Teflon. The friction coefficient of the material is high enough to ensure the transverse dynamic stability of the pump and motor assembly without requiring a very large force for centering it (Figure 3.8).

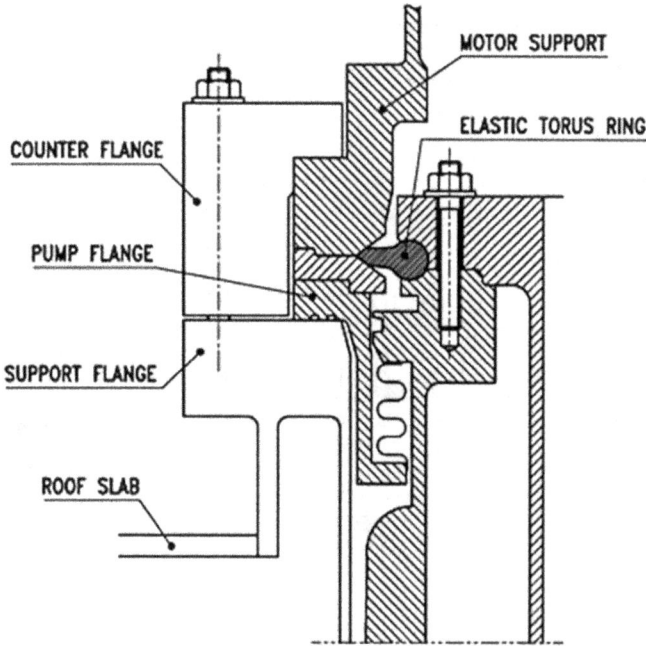

Figure 3.9 Support arrangement on roof slab for primary pump of SPX-I [15]. (Reproduced from R. M. Chabassier, J. J. Balteyron, J. F. Roumailhac, From Phenix to SuperPhénix: Mechanical Structures Assuring Reactor Vessel Tightness at Main Sodium Pump Penetrations, F2/6, International Conference on Structural Mechanics in Reactor Technology (SMiRT 4), San Francisco, USA. 1977).

However, the Teflon base allows the sliding plate, along with the pump support flange, to move laterally, thus absorbing the differential expansion arising from the movement of the pipe connected with the diagrid plenum below (at higher temperature) vis-a-vis the upper cooler support flange of the pump.

In the SuperPhénix-1 (SPX-1) primary pump [15], the pump flange is tightened against the support flange that is welded to the roof slab by a counter flange bolted to the support flange (Figure 3.9). The motor support is rigidly fixed to the pump flange. An elastic torus ring with an inner gear tooth profile permits ball and socket-type motion of the pump without stressing the internal structures of the reactor.

In the Prototype Fast breeder Reactor (PFBR) primary pump, upper compliant support is provided to accommodate the rotation of the pump assembly resulting from the radial thermal expansion of the reactor diagrid piping connected to the pump discharge nozzle while the axial thermal expansion of the pump assembly is accommodated in the connection (aka Pump to Pipe Connection PPC) between the pump discharge pipe and the reactor (core) diagrid pipe itself.

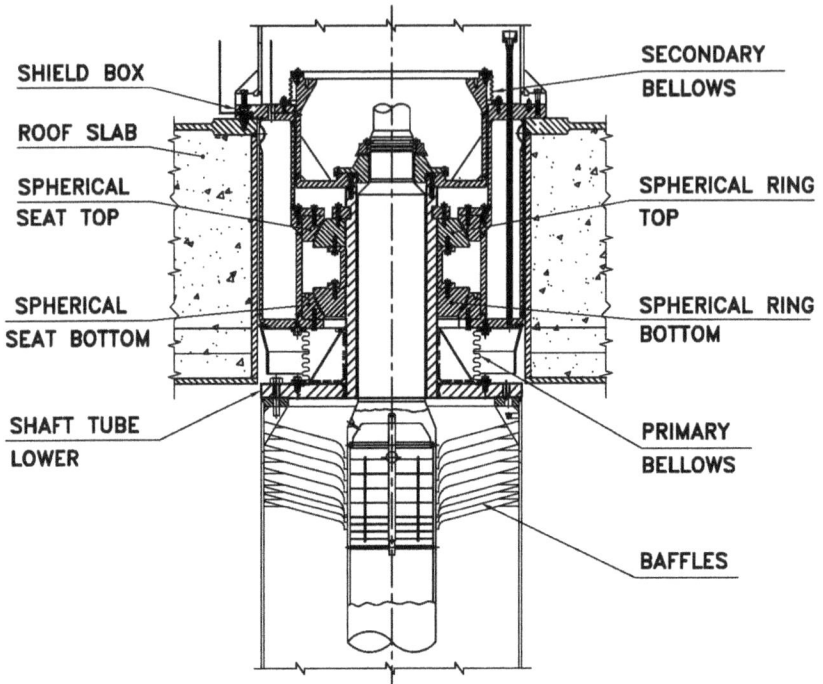

Figure 3.10 Spherical seat support to permit tilting of primary pump in PFBR.

The upper compliant support consists of a plain spherical bearing, which bears the weight of the pump assembly through a shaft tube and allows the rotor assembly (but not the shield plug) to tilt (maximum of 0.4°) in a vertical plane depending on the exact radial and axial temperature differential in the pump. The plain spherical bearing, in turn, through its connection with the shield plug, transfers the weight of the pump assembly to the upper flange support fixed to the roof slab. The spherical bearing housing, the shield plug, and the upper support flange are fixed in space and do not undergo tilting. The pitch circle diameters of the pump penetration in the roof and the bottom pipe receptacle have been chosen to have the pump aligned vertically at the nominal operating inlet temperature of the reactor (Figure 3.10).

Development of this critical component involved tests on a representative shaft with a dummy impeller to demonstrate the operation of the inclined rotating assembly pivoted about the plain spherical bearing. The assembly, supported at the bottom by an externally pressurized hydrostatic bearing and at the top by antifriction bearings, was operated at various inclinations expected in the reactor over the expected operating speeds and qualified. The prototype pump was tested at different inclinations during water tests to confirm the operability of the support arrangement.

In the SuperPhénix-2 (SPX-2) primary pump, the proposed design consisted of a rigid top support with mechanical seals for leak tightness, thus eliminating the tilting of the pump and thereby simplifying the design. The differential thermal expansion was accommodated by providing a sleeve at the pump discharge nozzle, suspended on gimbals from the pump diffuser with seal rings at both ends, that was free to move axially and tilt.

3.2.5.2 Pump discharge to reactor diagrid pipe connection (aka pump to pipe connection)

The pump discharge pipe is not permanently welded to the reactor diagrid piping to facilitate removal and re-insertion of the pump in the event of a maintenance requirement. This connection, together with the pump top support, accommodates the differential thermal expansion resulting from the top support being at room temperature and the bottom of the pump at the cold pool temperature ($\sim 400°C$). The design of this joint is as critical as that of the pump top support because it must:

(i) be capable of withstanding the reaction force resulting from thermal expansion and the resulting tilt of the pump – the pump will be vertical at the operating temperature corresponding to the rated power and inclined at all other conditions with the maximum tilt from the vertical being at room temperature;

(ii) keep the leakage flow rate from the pump discharge to the cold pool minimal and avoid cavitation;

(iii) facilitate relative motion due to thermal expansion and prevent jamming under thermal shock; and

(iv) permit vibration-free operation and be resistant to seismic motion. Some of the design concepts employed are as below.

For example, the EBR-II primary pump is connected to the primary circuit through a spring-loaded, ball-joint connector so that the pump in its normal position has the ball and socket held forcefully together by a force of 2720 kg (Figure 3.11). This design ensures minimal bypass of sodium (<0.2% of pump discharge) from the high-pressure pump discharge pipe to the low-pressure reactor pool. The joint also facilitates quick engagement of the pump discharge pipe with the diagrid piping during assembly and easy disengagement during removal. The arrangement accommodates differential thermal expansion between the connected parts, and the bellows compensates for any misalignment between the pump discharge pipe and the system piping.

In the EFAPP primary pump (Figure 3.12), the bypass of high-pressure liquid from the pump discharge pipe to pump suction is prevented by a seal held in position by the seal holder/seal face guide. The bellows permits relative expansion between the pump discharge pipe and the circuit piping.

In the Phenix primary pump [14], the pump discharge pipeline is connected to the reactor diagrid piping using a jointed collar/tilting connector that tilts and accommodates the differential axial and radial thermal expansion at the joint (Figure 3.13). The joint is designed to minimize the leakage

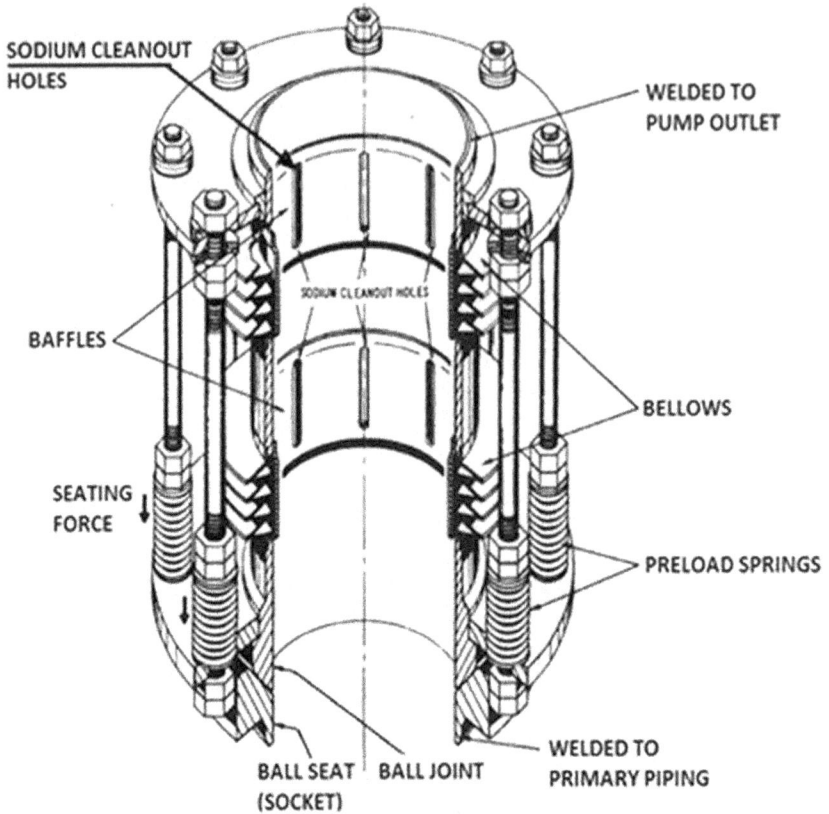

Figure 3.11 Ball-joint connector that connects pump discharge pipe to primary circuit in EBR-II Primary Pump [16]. (Reproduced with kind permission of Argonne National Laboratory, managed and operated by UChicago, Argonne, LLC, for the U.S. Department of Energy under Contract No. DE-AC02-06CH11357).

of sodium from the high-pressure pump discharge pipe into the relatively lower pump suction pool while permitting a certain degree of axial and lateral displacement without undue stresses on the pump and reactor piping while permitting easy insertion and removal of the pump assembly.

Figure 3.14 shows the arrangement in the primary sodium pump of PFBR [18]. Here the pump discharge nozzle engages with the piping connected to the reactor diagrid at the pump pipe connection (PPC). The PPC consists of two concentric annular sleeves between (a) the inner shell of the pump discharge nozzle and the pipe receptacle inner sleeve, and (b) the outer shell of the pump discharge nozzle and the pipe receptacle outer sleeve. A labyrinth restricts the leakage between the inner shell of the pump discharge nozzle and the pipe receptacle inner sleeve. In contrast, a compressed metal seal ring restricts

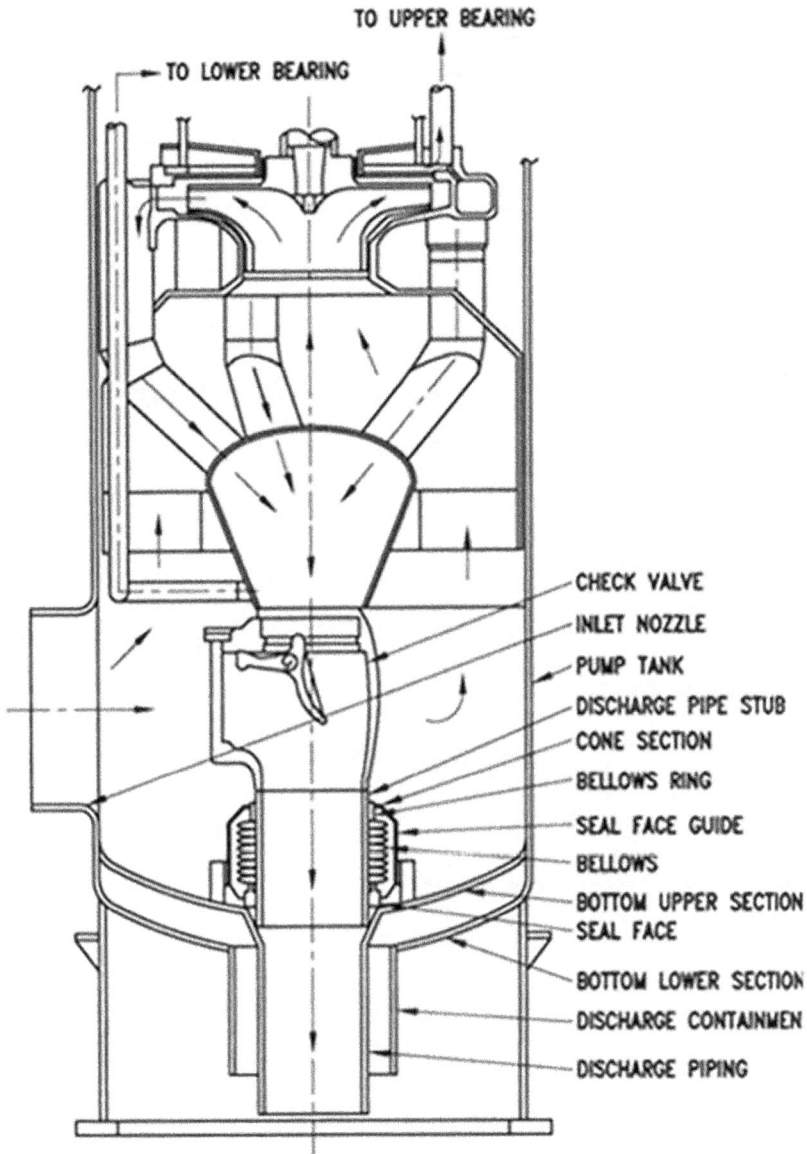

Figure 3.12 Arrangement to connect pump discharge to primary piping in EFAPP primary pump [17]. (Reproduced from D. J. Kniley, W. J. Carlson, E. Ferguson, O. G. Jenkins, Mechanical Elements Operating in Sodium and Other Alkali Metals, Vol II, Experience Survey, Report no. LMEC-68-5, June 1970).

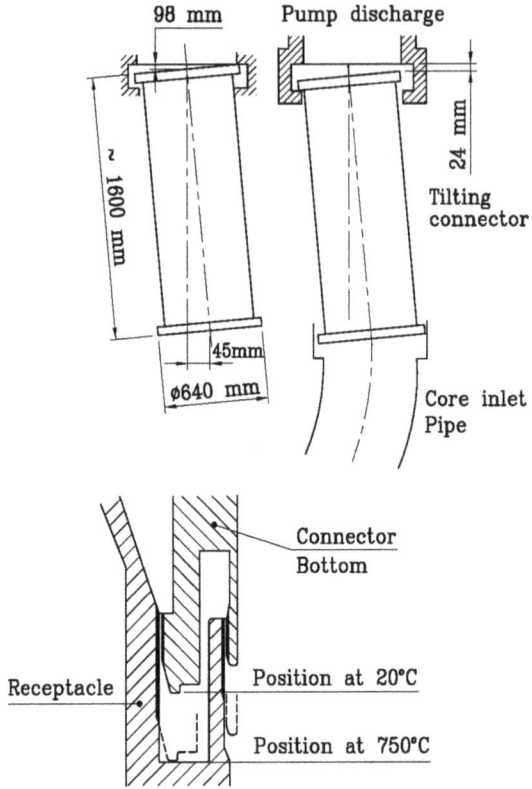

Figure 3.13 Connection between pump discharge pipe to core diagrid pipe in Phenix [14]. (Reproduced with kind permission of the International Atomic Energy Agency (IAEA)).

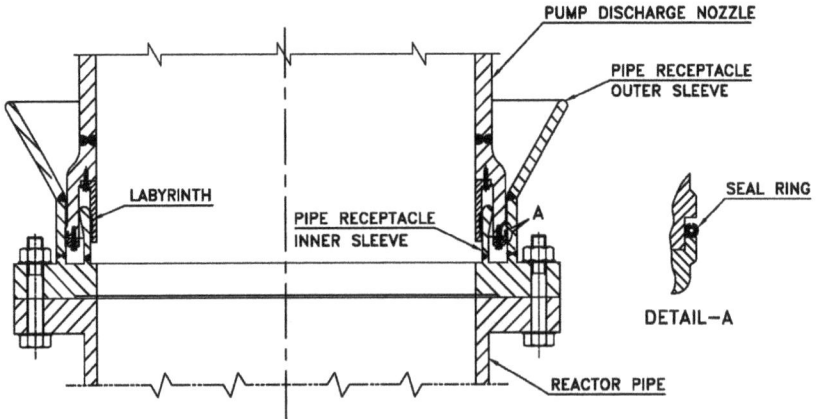

Figure 3.14 Pump discharge nozzle to reactor pipe connection in PFBR primary pump.

leakage between the outer shell of the pump discharge nozzle and the pipe receptacle's outer sleeve. The seal ring also prevents looseness of the joint while permitting easy assembly. The outer sleeve connected to the pipe receptacle has a funnel-shaped entry whose inner diameter is governed by:

(a) the expected lateral thermal expansion at PPC;
(b) the maximum inclination allowed at the upper compliant support; and
(c) the maximum permissible eccentricity due to the roof slab penetration clearance.

The conical funnel entry facilitates the centring of the pump during insertion into the reactor. The seal ring and the labyrinth restrict the sodium leak rate into the cold pool, which is less than 1% of the pump discharge.

During the development of the SPX-I primary pump, as many as four design variants were tested to finalise the pump discharge to diagrid pipe connection (aka LIPOSO[2]). The initial design consisted of a thin shell (3 mm) that expanded under the high pressure of the pump discharge and was forced against the stationary part to limit the leakage flow rate. The full-scale testing of the design and its variants of increased thickness resulted in noise and vibration. A rigid labyrinth joint with radial clearance of 0.8 mm–0.9 mm was finally selected after confirmation of cavitation free operation of a full-scale prototype in sodium. An alternative design consisting of a split packing ring type of seal that expanded under the high pressure of the pump discharge and provided good sealing between the removable and stationary parts without hydraulic instability was selected as the backup solution [19].

The joint between the pump discharge pipe and the reactor diagrid pipe, in the concept used in SPX-2 [15], is shown in Figure 3.15. The pump discharge nozzle consists of external and internal cylindrical sleeves. The external sleeve (1) acts as a guide, while the inner sleeve is provided with an elastic, hard plated surface. This surface mates with the hard plated sleeve (3) to form a semi rigid, leak tight joint that influences the critical speed of the rotor assembly as well as the natural frequency of the pump assembly. The leak-arresting contact surfaces of the external and internal sleeves are spherical, thus permitting universal type movement around the torus.

The parts mating with these sleeves consists of a hard plated sleeve (3), the female part of the leak-arresting joint, and a non-plated conical receptacle (4). This conical receptacle has a push fit with the pump and facilitates the centring of the pump during assembly. It is possible to remotely dismantle both these parts from the top of the reactor vessel after the removal of the pump. The assembly of (3) and (4) is fixed to the reactor diagrid pipe but can be remotely dismantled, if required. A sleeve (6) provides leak tightness between the reactor diagrid piping (5) and this assembly.

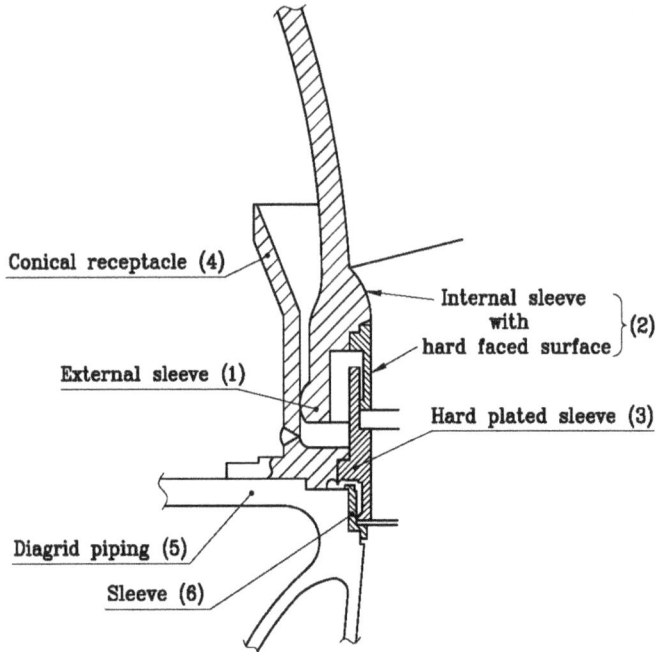

Figure 3.15 Pump discharge pipe to reactor diagrid pipe connection in SPX-2 [15].
(Reproduced from R.M. Chabassier, J. J. Balteyron, J. F. Roumailhac,
From Phenix to SuperPhénix: Mechanical Structures Assuring
Reactor Vessel Tightness at Main Sodium Pump Penetrations, F2/6,
International Conference on Structural Mechanics in Reactor
Technology (SMiRT 4), San Francisco, USA, 1977).

3.2.6 Other design features

3.2.6.1 Preventing gas entrainment

In the case of pumps that draw suction from a sump (as opposed to pumps
with piped suction), there is the possibility of vortex and gas entrainment in
pump suction if the submergence is insufficient. Gas entrainment can be prevented by providing an antivortex plate in pump suction as in the primary
pumps of RAPSODIE and FBTR.

The BN-350 primary pump provides a detailed system to prevent gas
entrainment [20]. The arrangement (Figure 3.16) consists of an overflow
tank that has a floating regulator that maintains the minimum level required
to prevent gas entrainment in the pump tank.

The overflow tank also helps to release any gas that may be entrained
along with the sodium. When the pump is started, the pressure drop in the
suction line causes the overflow tank's float valve to move down, close
the tank outlet, and stop further reduction in level. When leakage from the

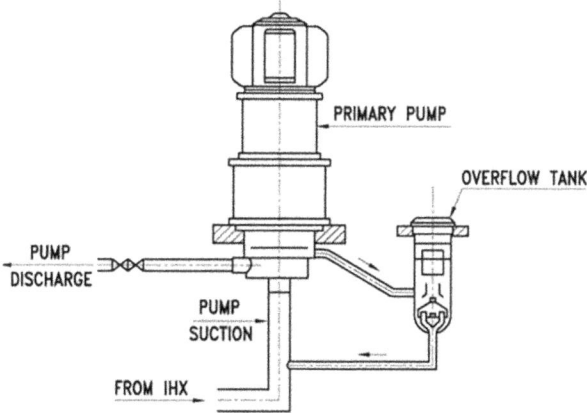

Figure 3.16 Arrangement to avoid gas entrainment in BN-350.

pump tank increases, the float valve rises until the level in the overflow tank is sufficient to match the outflow with the inflow. This system has been successfully used.

In the Prototype fast reactor (PFR), space was provided between the pump outlet plenum and the non-return valve to include a gas separation device. The device consisted of fixed vanes that caused the passing liquid to rotate, resulting in gas separation and removal from the vortex's centre through a central pipe to the argon cover gas space. The device, however, was not installed, and served only as a standby for use, if required [13].

3.2.6.2 Thermal baffles in cover gas space

The heat transferred from the free surface of sodium by conduction through the shaft, convection through the argon cover gas, and radiation increases the heat load on the cooling oil circulating through the top bearings and seals. Thermal baffles in the pump cover gas space reduce the heat load to the bearings and seals and are also effective in preventing the formation of convective currents from stabilising in the cover gas space and causing asymmetric temperature differential in the pump tank. This phenomenon was observed in the JOYO primary pump where distortion of the pump inner tank, due to a circumferential temperature difference of 50°C, was rectified by installing thermal baffles in the cover gas space. Similarly, during the preheating, in preparation for sodium testing of the FFTF primary pump [21], convection currents resulted in shaft bending and an increase in the breakaway torque of the shaft. The introduction of baffles in the space between the pump tank and the shield plug resolved the problem by preventing any temperature differential from arising due to the stabilisation of convection currents.

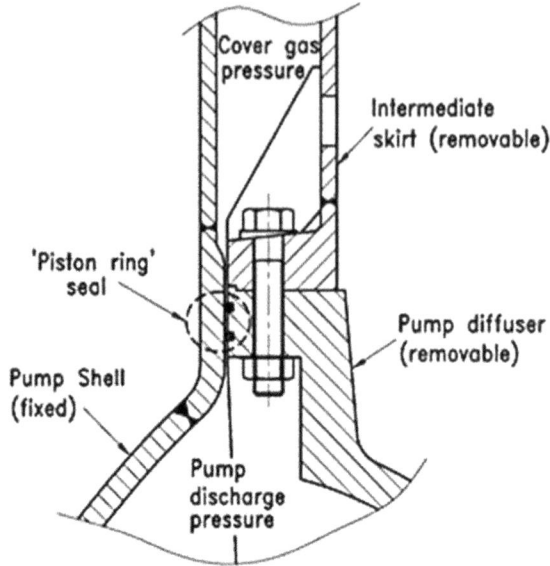

Figure 3.17 'Piston ring' seal to separate high- and low-pressure regions in pump.

3.2.6.3 Sealing

Mechanical seals are the most common devices used to maintain leak tightness in the cover gas space. Double and triple mechanical seals have been employed for this purpose. A repair seal is also provided to maintain the leak tightness of the pump cover gas space during the maintenance of top bearings and seals. In the PFBR pumps, the repair seal consists of a pair of O-rings mounted on the stationary surface. During the dismantling of the top bearing and seals, the shaft is lowered until the oil deflector plate mounted on the shaft bears down on the double O-ring to seal the pump cover gas space from the atmosphere.

In piped suction pumps where the pump discharge is connected to the pump tank, an additional sealing arrangement is critical to prevent liquid leakage from the high-pressure side (pump tank/bowl) to the suction piping or cover gas region. A high-pressure seal separates the pump discharge region from the pump cover gas space in such systems. For example, in the secondary pump of PFBR, a piston ring seal prevents leakage from the discharge bowl to the free sodium level in the cover gas space, and a labyrinth prevents leakage from the bowl to the suction piping. Figure 3.17 shows the piston ring seal used in the PFBR secondary pump. The leakage flow rate through the joint is negligible compared to the pump's design flow rate [22].

3.3 SUMMARY

This chapter has discussed the options to be considered in the hydraulic and mechanical design of a centrifugal sodium pump. Selection of the appropriate design alternative has a direct bearing on material selection, pump sizing, and capital cost, as well as operating reliability, ease of maintenance, and operating costs.

NOTES

1 This pump, however, had a rigid coupling between the pump shaft and motor shaft which resulted in alignment problems.
2 LIPOSO is the French acronym for Liaison Pompe-sommier, meaning Pump-diagrid connection.

REFERENCES

1. S. Lazarkiewicz and A.T. Troskolanski, *Impeller Pumps*, Pergamon Press, 1965.
2. A.J. Stepanoff, *Centrifugal and Axial Flow Pumps*, Wiley, 1975
3. J.F. Gulich, *Centrifugal Pumps*, Springer, 2008
4. K.M. Srinivasan, *Rotodynamic Pumps* (centrifugal and axial), New Age International Private Limited, 2008.
5. I.J. Karassik, J.P. Messina, P. Cooper and C.C. Heald, *Pump Handbook*, 3rd edition, McGraw Hill, 2001
6. A.H. Church, *Centrifugal Pumps and Blowers*, Krieger Publishing Company, 1972
7. Addison, *Centrifugal and Other Rotodynamic Pumps*, Chapman and Hall, 1966.
8. V.S. Lobanoff and R.R. Ross, *Centrifugal Pumps: Design and Application*, 2nd edition, Gulf Publishing Company, Houston, Texas.
9. V.M. Cherkassky, *Pumps, Fans and Compressors*, Mir Publishers, Moscow, 1985.
10. A.G. Salisbury, Current Concepts in Centrifugal Pump Hydraulic Design, Paper no. C177/82, IMechE Conference Publications 1982-11.
11. W. Babcock, State of Technology Study - Pumps Experience with High Temperature Sodium Pumps in Nuclear Reactor Service and their Application to F.F.T.F., BNWL-1049, December 1969.
12. Fast Reactor Database 2006 update, No. IAEA-TECDOC-1531, International Atomic Energy Agency, December 2006.
13. J.M. Laithwite and A.F. Taylor, Hydraulic Problems in the P.F.R. Coolant Circuit, *Proceedings of a Symposium on Progress in Sodium Cooled Fast Reactor Engineering*, Monaco, March 1970.
14. W. Raozynski, J.P. Delisle, and J.L. Befree, L'Evolution des Pompes a Sodium de Rapsodie a Phenix et a la Filiere, *Proceedings of a Symposium on Progress in Sodium Cooled Fast Reactor Engineering held by the International Atomic Energy Agency in Monaco*, 23–27 March 1970.
15. R.M. Chabassier, J.J. Balteyron and J.F. Roumailhac, from Phenix to SuperPhénix: Mechanical Structures Assuring Reactor Vessel Tightness at Main Sodium Pump Penetrations, F2/6, *International Conference on Structural Mechanics in Reactor Technology (SMiRT 4)*, San Francisco. 1977.
16. J.R. Davis, G.E. Deegan, J.D. Leman and W.H. Perry, Operating Experience with Sodium Pumps at EBR-II, Report no. ANL/EBR-027, Argonne National Laboratory, October 1970.

17. D.J. Kniley, W.J. Carlson, E. Ferguson and O.G. Jenkins, Mechanical Elements Operating in Sodium and Other Alkali Metals, Vol II, Experience Survey, Report no. LMEC-68-5, June 1970.
18. K.V. Sreedharan, S. Athmalingam, V. Balasubramaniyan, A.S.L.K. Rao, P. Chellapandi, S.C. Chetal, Design and Manufacture of Sodium Pumps of 500 MWe P.F.B.R., I.N.S. *Annual Conference (INSAC-2009)*, January 2010, Chennai, India.
19. J. Guidez and G. Prele, Superphénix: Technical and Scientific Achievements, Atlantis Press, 2017.
20. F.M. Mitenkov, E.G. Novinsky, V.M. Budov, *Main Circulation Pumps for Atomic Power Stations*, edited by F.M. Mitenkov, member, Academy of Sciences, U.S.S.R., 2nd Edition ENERGOATOMIZDAT, Moscow, 1990.
21. R.W. Atz and M.J. Tessier, High Temperature Testing of a Sodium Pump, Presented at *A.S.M.E. Winter Annual Meeting*, December 1978, A.S.M.E. Publication No. 78WA/NE12, 1978.
22. C.K. Keshava Kumar, K.V. Sreedharan, A.S.L.K. Rao and S.C. Chetal, Design and Selection of Slip Joint for Sodium Pump, *32nd National Conference on Fluid Mechanics & Fluid Power*, December 2005, Osmanabad, India.

Chapter 4

Manufacture of centrifugal sodium pumps

4.1 INTRODUCTION

Many components in a sodium centrifugal pump require development efforts in their manufacture. This requirement arises from the stringent dimensional tolerances specified for fabrication and machining and the need to maintain dimensional stability during operation at high temperatures for extended periods.

The components that demand special attention during the manufacture of a sodium pump are:

1. Pump impeller, diffuser, and suction casing.
2. Pump shaft.
3. Hydrostatic bearing.
4. Pump top support (such as spherical seat support for the PFBR primary pump).
5. Non-return valve at the pump discharge.

In the following sections, we shall discuss the technical requirements and challenges in manufacturing the above components with some examples.

The material of construction of pump components must meet several requirements; some contradictory to others. These include good mechanical strength (in cold and hot conditions), good ductility, impact strength and hardness, good weldability, and high resistance to creep.

Table 4.1 lists the major components and the typical materials used for the construction of cold leg pumps.

4.2 COMPONENTS MANUFACTURE REQUIREMENTS

4.2.1 Pump impeller, diffuser and suction/discharge casings

Castings for nuclear reactor applications are purchased based on detailed specifications, testing procedures, and certification to ensure high quality.

Table 4.1 Typical materials used for pump components

Part	Material	Product form
Hydraulic parts such as impeller, diffuser, suction/discharge casings	Austenitic stainless steel 304 LN[a]	Casting (ASTM A 351 Grade CF3)
Other structural components that are in contact with sodium or its vapour e.g., pump vessel, thermal baffles, antivortex plate, seal, and bearing housing etc.	Austenitic stainless steel 304 LN[a]	Plate
Pump shaft, sodium bearing, and supports, wearing rings, etc.	SS 304 LN[a]	Forging
Components not in contact with sodium or its vapour e.g., Housing for thrust bearing and mechanical seals etc.	Fine grained carbon steel with high impact strength	Plates/forgings (e.g., SA 516, Gr. 65)
Fasteners (bolts and nuts) in contact with sodium or its vapour		Rounds/bars (e.g., SA 453 / SA 194)
Fly wheel	Closely controlled carbon steel with high fracture toughness and low impurity content	Forging/Plates (e.g., SA-508 Grade 2)
Hard facing materials	Colmonoy	Filler wire/rod/ powder

[a] The material will be SS 316 L or SS 316LN for hot leg pumps. Shafts, in particular, have been made of SS 316 or Titanium stabilised austenitic stainless steel.

The specifications are usually based on codes for nuclear components such as ASME Sec III or RCC-MR (French code for design and construction of fast reactors). Either the principal contractor or the end user specifies all technical parameters for supply of the equipment, such as pumping medium, operating temperature, flow rate, delivery pressure, operating speed, power, and efficiency.

Castings (ASTM A 351 Grade CF3 for cold leg installation and ASTM A 351 Grade CF3M for hot leg installation) are used for components such as the impeller, diffuser/volute casing, and suction casing. They must meet the specified chemical composition and mechanical properties in addition to certain delta ferrite content for better weldability.

Before undertaking manufacture, a detailed Quality Assurance Plan (QAP) and Process Plan are prepared, and various procedures related to raw material inspection, manufacturing, and product inspection are approved.

Samples are tested for chemical composition and mechanical properties as well as for estimation of ferrite content (weldability requirement). The following main procedures need to be approved by the user:

- Heat treatment of castings.
- Radiography method and acceptance criteria.
- Liquid Penetrant Examination with acceptance criteria.
- Pressure test wherever needed.
- Repair welding procedures and limitations.
- Ultrasonic testing as supplementary, where useful.
- Dimensional inspection.

The purchaser approves the drawings of patterns before starting the work and inspects the patterns before making cores and moulds. Fine zircon sand having high-temperature stability is recommended to ensure a good surface finish and minimise inclusions, especially for inaccessible regions such as vanes, interiors of shrouds, and other similar areas. The casting technique should be of such a high standard that Radiographic Quality Level 2 should be realised at the metal pouring stage and without excessive welding repairs [1]. The castings are inspected using Liquid Penetrant Examination (LPE) technique, and unacceptable surface defects, if any, are removed by grinding or machining.

Minor repairs by welding are permitted. Those exceeding the specified limit are inspected by radiography and documented. A limit is also specified on the total area of inclusions, including that resulting from slag.

Radiographic examination of castings shall be done after rough machining. Based on the requirement, all sections shall be radiographed using X-ray or gamma ray.

The examination is in accordance with recommended practice (e.g., ASTM E 94), and the product is to satisfy the specified acceptance level (e.g., Severity Level 2 of ASTM E446) for shrinkages, slag inclusions, and gas porosity. Cracks, hot tears, mottling, etc., are not acceptable.

Complete and detailed shooting sketches and marking of coordinates on casting surfaces are recommended for a thorough radiographic examination. This ensures complete coverage by radiographic films and enables precise and easy defect location.

All parts are subjected to solution annealing heat treatment after repair welding. The heating rate, soaking temperature, soaking time, permitted temperature variation on the job, and cooling rate are specified, and a record is maintained. The castings of rotating parts (e.g., impeller) are dynamically balanced to the required grade after final machining.

Pickling and passivation of the castings shall be completed before final packing and dispatch.

The complete documentation, as agreed before ordering, shall be furnished to the customer along with the finished component.

4.2.2 Pump shaft

A critical component of the pump equal in importance to the pump hydraulics is the pump shaft. A reactor sodium pump shaft usually consists of several forged pieces (typically two solid ends and a central hollow portion) welded together. The solid bottom end is employed to mount the impeller and the journal of the hydrostatic bearing, while the top end is suited to mount the top radial and thrust bearings and the mechanical seals. The balance welding technique is employed to integrate the parts because no machining of the hollow portion or straightening of the shaft after welding is permitted. The hollow central portion of the shaft is evacuated and provided with a set of thermal baffles at its top (for large-diameter hollow shafts) to prevent convection inside the shaft and reduce radiation heat transfer to the top of the shaft).

The following requirements are to be satisfied while procuring the forging:

- The forgings shall be in solution heat-treated condition along with documents on the heat treatment cycle.
- The material shall be in rough machined condition for further fabrication of the shaft assembly at the pump supplier's or subcontractor's works.
- The material certificates furnished include those on chemical composition, tensile test (at room temperature and operating temperature), impact test, intergranular corrosion test (e.g., as per ASTM A262), inclusion content, grain size, and delta ferrite content.
- Certificates on examination of forgings furnished include those on visual examination, LPE, and ultrasonic examination to eliminate surface or internal defects.

The following are the significant steps in manufacturing:

(a) Machining: The solid ends are rough machined with sufficient allowance on radius and various section lengths for finish machining. The hollow central portion is finish machined with adequate allowance on length for anticipated weld shrinkage. All dimensional inspection is done at 293K, and final machining is done in an enclosed space maintained at a controlled temperature.

(b) Welding: Welding is a critical activity in the manufacture of the shaft. Achieving defect-free welds with minimal distortion is challenging because the central hollow portion is finish-machined before it is welded to solid ends on either side. Welding of the solid and hollow portions is done with TIG welding using filler wires. Welding procedure qualification and performance qualification are done with a portion of the extra length provided in the hollow portion. A joint design with no root gap and a backing cum centring ring is preferred for the two circumferential welds between the hollow and solid portions of the shaft to improve centring and reduce distortion during welding. Figure 4.1 shows a shaft manufactured as part of the pump development work at IGCAR [2]. The typical weld geometry between the solid and hollow portion is shown in the detail. Balance welding technique and control of welding parameters and sequence are used to achieve minimum distortion. The shaft assembly is monitored for distortion during the welding process with the help of several dial gauges. Radiography of the joints is done after root pass, final pass, and also at an intermediate stage after completing ~50% thickness, depending on the thickness of the hollow part. This step allows for ease of repair midway, if needed.

(c) Heat treatment: The shaft is heat-treated in vertical condition to relieve the stresses generated during rough machining and welding. The heat treatment is done in an electrically heated furnace in a reducing atmosphere of argon with 10% hydrogen to avoid scaling/oxidation of the shaft surface. A schematic of the shaft in the furnace during heat treatment is shown in Figure 4.2.

The following conditions are specified for the heat treatment process:
- Heating rate from room temperature to soaking temperature.
- Soaking temperature and duration.
- Maximum permitted temperature gradient on the shaft.
- Permitted temperature variation along the entire length of the shaft.
- Maximum cooling rate in the furnace.

(d) Finish machining and dimensional inspection: Finish machining, threading (e.g., lock nut), and hole drilling in the solid end (for hydrostatic bearing feed) are followed by helium leak testing and evacuation of the hollow central portion of the shaft, after which dimensional inspection is completed.

(e) Dynamic balancing: The shaft and the rotating parts, such as the impeller, hydrostatic bearing journal, sleeves, and half coupling, are individually balanced. The integrated rotating assembly is then balanced to a high grade of balancing as used in gas turbine rotors that operate at high temperatures. (e.g., ISO Grade 1.0 or better).

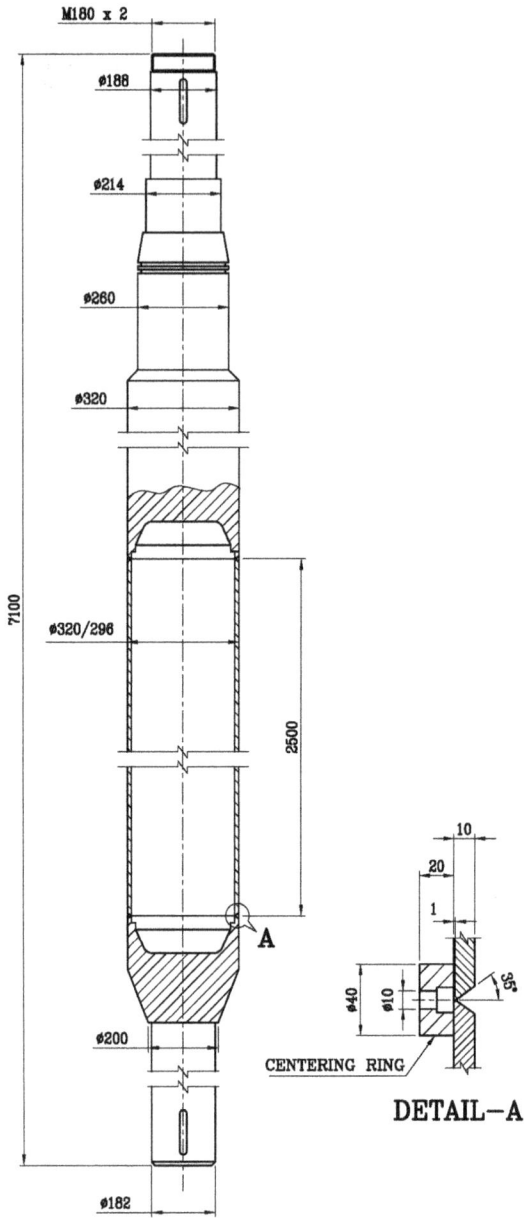

Figure 4.1 Typical shaft of a sodium pump.

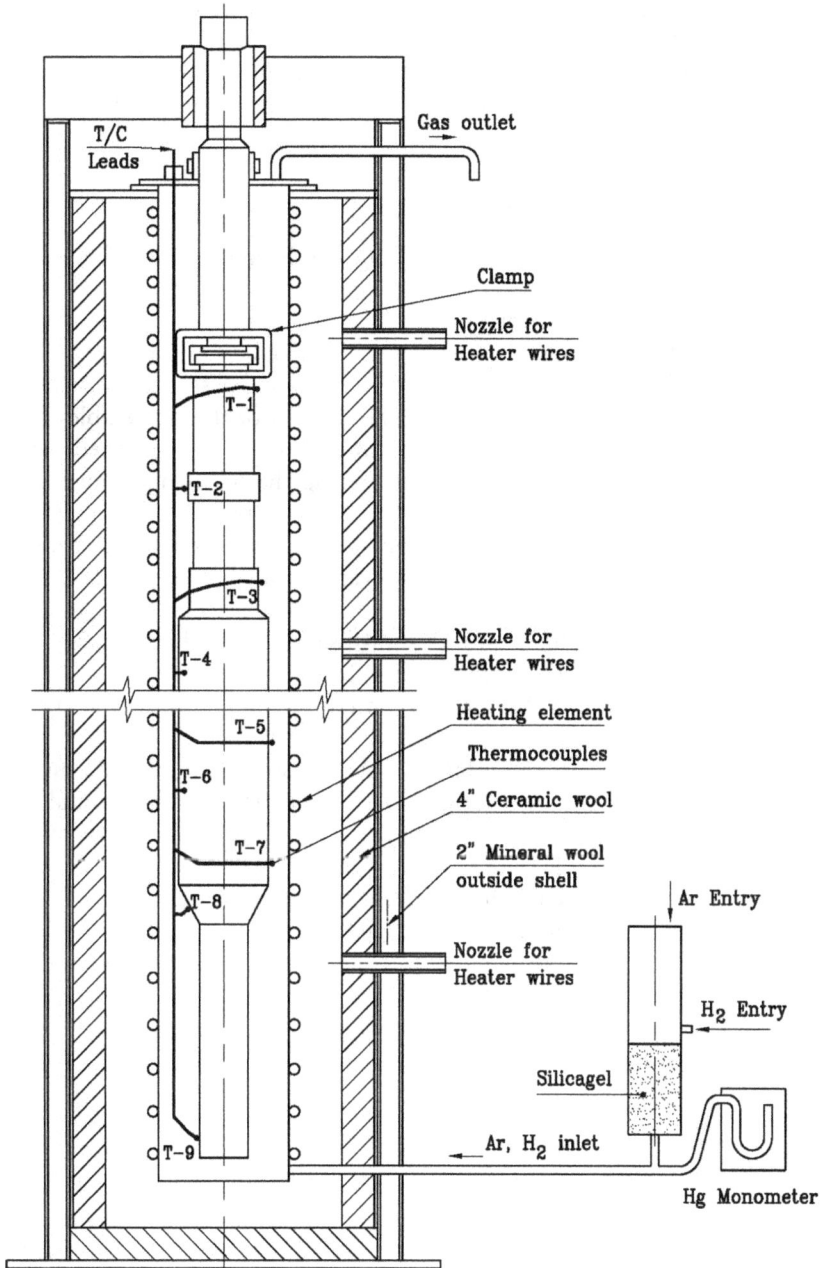

Figure 4.2 Schematic of heat treatment of shaft in furnace.

4.3 MANUFACTURING EXPERIENCE OF FBTR PUMPS [3]

The FBTR pumps are described in Section 2.2.2. The following paragraphs outline the challenges in manufacturing critical components of the pump.

4.3.1 Fixed vessel of primary pump

The removable assembly of the primary pump is located inside a fixed vessel of height 4605 mm and maximum outer diameter of 1080 mm. The shell thickness of the vessel is 6 mm along the length, except at the bottom, where it is 10 mm. The top flange of the fixed vessel is bolted to the casing of the biological shield. The flange of the removable pump assembly is bolted to the top flange of the fixed vessel, and the bottom of the removable pump assembly is aligned in place by a guide at the bottom of the fixed vessel. Therefore, the arrangement required achieving the following dimensional tolerances to ensure satisfactory operation of the pump.

Perpendicularity of the top and bottom faces of the vessel top flange, with respect to the vessel axis, within 0.1 mm/m and parallelity of the faces within 0.075 mm.

Concentricity within 0.5 mm of the diffuser guide of diameter 724 mm with respect to the vessel bottom guide of 230 mm.

Machining tolerance of 0.0 mm/0.01 mm on pump guide bore of diameter 250 mm.

Machining tolerance of 0.0 mm/2.0 mm on vessel length of 4605 mm.

Machining of the vessel was done in a single setting on a horizontal boring machine, having positional accuracy of 0.001 mm, with the vessel mounted on a rotary table (Figure 4.3) having an accuracy of 5 s. The vessel axis was aligned with the machine spindle axis optically.

High vibrations experienced during machining necessitated modification of the vessel supports and clamping arrangements before the vessel was machined to the required tolerances.

The bottom face of the flange was machined in one setting using special fixtures.

4.3.2 Shaft of primary pump

The shaft is of composite geometry with top and bottom solid ends and a central hollow portion. Solid and hollow forgings of Z20CNW 22 are used. In operation, the shaft is supported on a radial hydrostatic bearing at the bottom and conical roller bearings at the top, with the span between the bearings being 3300 mm. The total length of the shaft is 3940 mm, and to ensure vibration-free operation, the concentricity over the entire length is

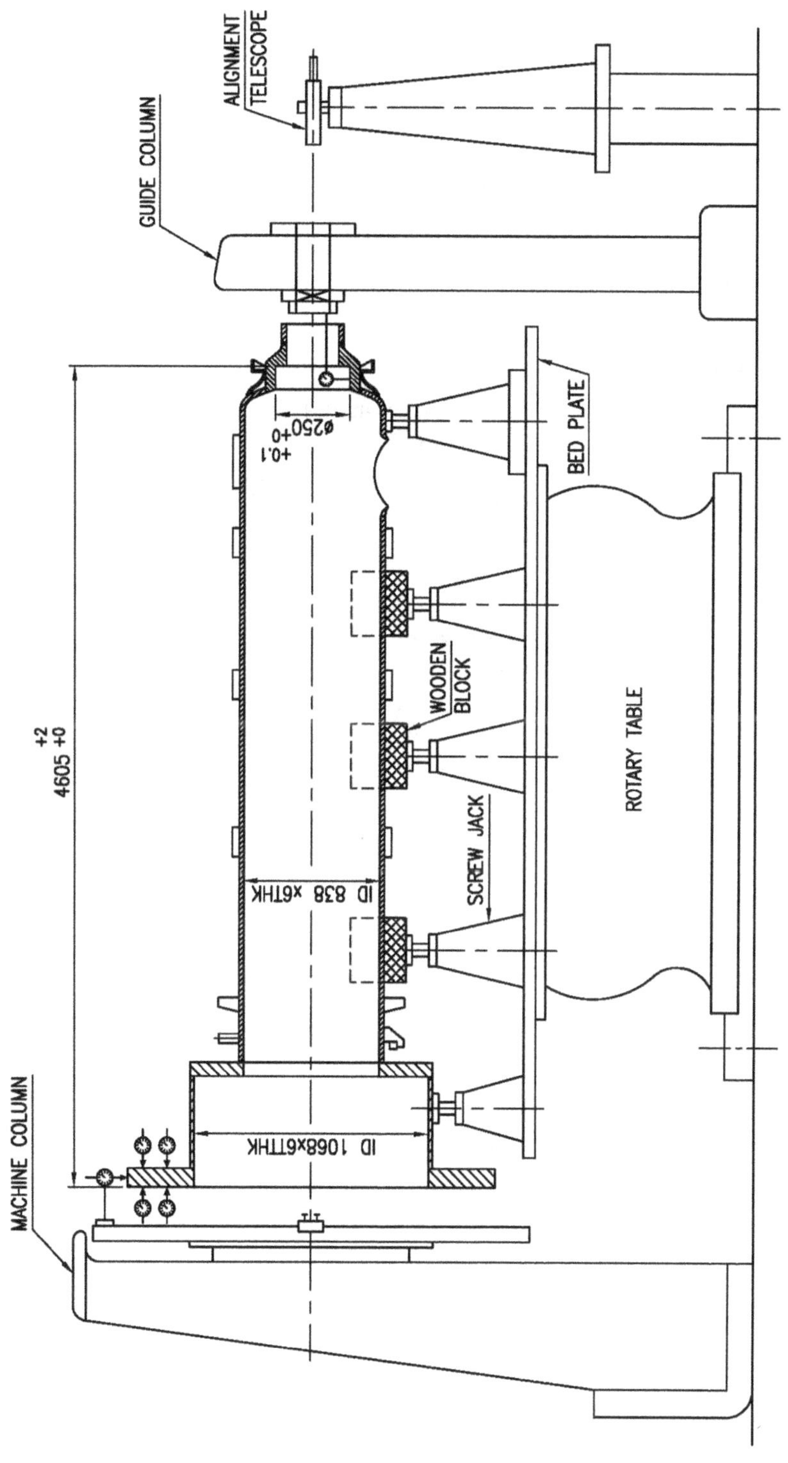

Figure 4.3 Arrangement for machining fixed vessel of primary pump.

limited to less than 0.01 mm. The required concentricity was achieved after final machining, and no straightening was allowed at any stage of manufacture. Balance welding technique was adopted with control of heat input, and weld shrinkage was limited to 7.7 mm as against the estimated value of 8 mm. The concentricity of the rough machined shaft was measured to be within 0.5 mm. Stress-relieving heat treatment was done in the vertical condition in reducing atmosphere (argon + 10% hydrogen). Final machining was done in temperature-controlled environment to achieve the specified concentricity over the entire length.

4.3.3 Impeller of primary pump

The impeller is of single-stage top suction design. It is made from casting Z8CND19-10 (equivalent to ASTM CF8M). Figure 4.4 is a sketch of the impeller showing the dimensional tolerances. The final machined diameters of the steps on the front and rear shrouds are to be concentric

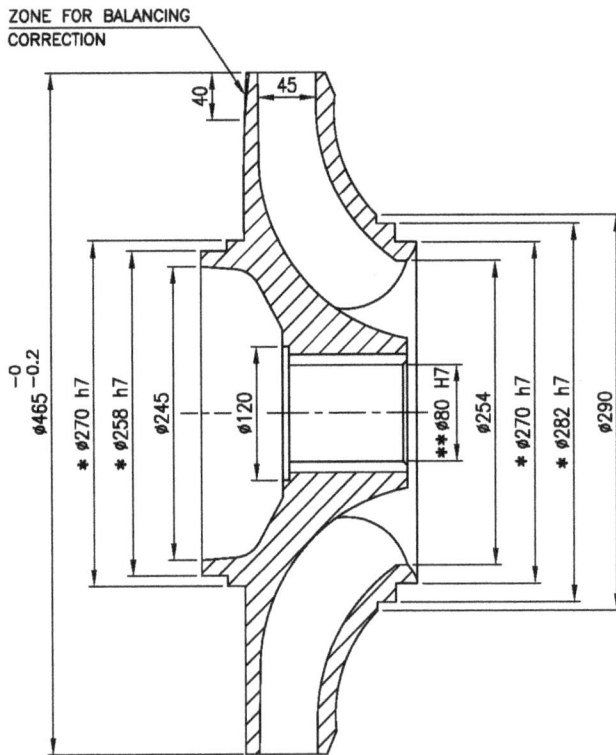

Figure 4.4 Sketch of impeller of primary pump with important tolerances.

with the bore of the impeller to within 0.01 mm. In the case of the first impeller, the machining of the 258h7, 270h7, and 282h7 steps was taken up after completing profile machining and finish machining of the 80H7 bore. A mandrel was inserted in the 80 mm diameter bore to hold the impeller during the machining of the steps. After completing the machining, the machinist committed an error while removing the mandrel resulting in damage to the bore. In order to rectify the mistake, the bore was enlarged to 81H7, and the corresponding mating shaft diameter was suitably modified.

The process plans of the other impellers were therefore modified to prevent the recurrence of a similar mistake by completing first the finish machining of the steps while holding the impeller using a mandrel in the undersized bore and finish machining the bore later on a jig boring machine after taking reference from the finish machined steps. The static balancing of the impeller was done twice during the rough machining stage to simplify the final balancing of the rotating assembly.

The impeller was subjected to DP examination and also to radiographic examination. Defects of types A, B, and C of severity Level 2 and below (as per ASTM E446) were detected. These were accepted as they were within limits specified in the technical specifications.

4.3.4 Expansion tank of secondary pump

The secondary sodium pump is housed in a spherical vessel that also serves as the expansion tank of the secondary circuit. The spherical shell is constructed from AISI 316 plates.

A major problem surfaced while welding the top flange to the top hemisphere and the bottom cone to the bottom hemisphere. In order to achieve the desired tolerances, sufficient machining allowances were provided on the top flange face of the upper hemisphere as well as the bottom cone of the lower hemisphere based on estimated weld shrinkages at the different joints. However, despite using fixtures, weld shrinkage of 7 mm was observed at the joint between the top flange and the upper hemisphere. The welding technique was reviewed, and detailed calculations showed that the overall height of the component could be reduced by an additional 14.5 mm due to weld shrinkages. It was considered possible to accommodate this reduction through allowances provided in the equator line joint and on the bottom cone of the lower hemisphere. In addition, the pump lower guide was not machined. However, due to oversight, the manufacturer machined the pump guide. Finally, the crucial total height of the spherical vessel was maintained by cutting the bottom cone and welding an intermediate conical piece between it and the pump guide (Figure 4.5).

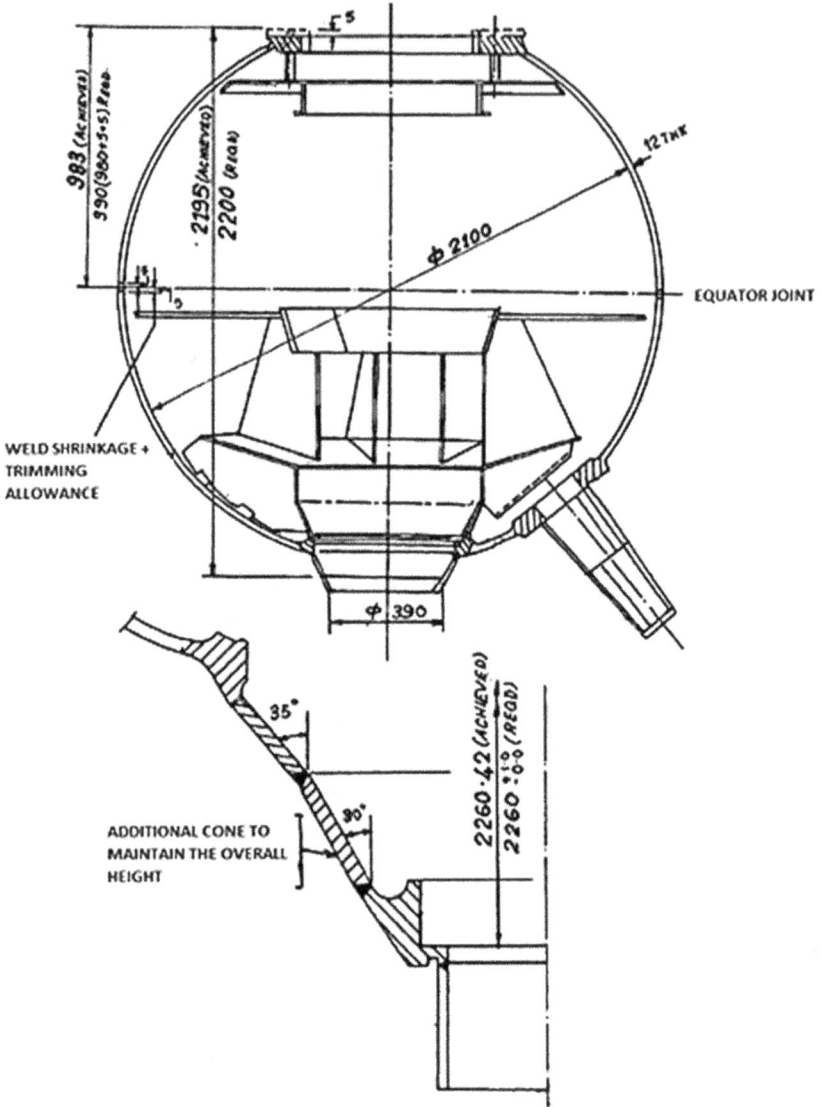

Figure 4.5 Expansion tank of secondary pump.

4.4 SUMMARY

Once the design and detailing of the components of a pump are completed, the next challenge is to manufacture the components to the desired tolerances. Manufacturing is a critical activity involving diverse tasks such as material qualification, fabrication, welding, machining, and quality control. The challenge lies in ensuring the specified tolerances for individual components and the final stacked-up tolerance of the assembly. Heat treatment is equally important to prevent distortion and ensure dimensional fidelity during long-term operation at high temperatures. This chapter focuses on the challenges encountered during the manufacturing of critical components for a centrifugal pump.

REFERENCES

1. F.W. Kamber, Steel castings for Nuclear Power Plant Pumps, Paper 93/74, *Proceedings of Institution of Mechanical Engineers*, 1974, U.K.
2. R.D. Kale, A.S.L.K. Rao, S. Baskar, K.V. Sreedharan and K. Balachander, *Mechanical Design Features of a High-Temperature Reactor Coolant Pump, All India Conference on Pumps*, Organised by The Institute of Energy management, Mumbai, India, April 25 & 26, 1997.
3. B.P.S. Rao, S. Ghosh, M. Mannaru and R. Chandramohan, Manufacture of F.B.T.R. Sodium Pumps, *Proc. of National Symposium on Pumping Equipment in Nuclear Industry and Thermal Power Plants*, February 24–25, 1994, B.A.R.C., Bombay, India.

Chapter 5

Centrifugal sodium pump testing

5.1 INTRODUCTION

The primary goals of pump testing are:

(i) To demonstrate to the customer the fulfilment of technical guarantees on hydraulic and mechanical performance. This applies to both pumps within the manufacturer's standard product range and to custom designs.
(ii) To demonstrate the satisfactory operation of the pump along with the site intake arrangement in cases where the system is sensitive to the intake flow path.

Tests on pumps are classified into:

1. Acceptance tests: These are conducted on the pump manufacturer's test bed, in the presence of the purchaser or his representative, to demonstrate the fulfilment of performance guarantees detailed in the purchase document. The manufacturer's facility is constructed and instrumented in conformance with established pump testing standards/codes, and the test results are highly accurate and reproducible.
2. Field tests: These tests are carried out in the field/plant. The pump is tested onsite in the service conditions expected during operation i.e., simulating the intake arrangement, operating conditions, pumped liquid and any other critical parameters. The accuracy of the tests will depend on the instrumentation employed and the accessibility at the site to deploy appropriate instruments.

    These test results can also be used to decide on acceptance/rejection if agreed to explicitly in the purchase document.
3. Model tests: These are high-accuracy tests, usually done on scaled-down models during the development of new/improved designs. The tests demand complete fidelity between the model and the prototype and is achieved by conforming to appropriate scaling laws and testing

DOI: 10.1201/9781003460350-5

standards. The accuracy of the instrumentation employed is also higher than that used in routine testing.

4. Periodic tests: These are tests scheduled in the field at regular intervals to detect changes or variations in performance resulting from wear and tear. The instruments employed are standardised and of unvarying accuracy to enable the identification of impending problems.

Mechanical sodium pumps are designed for reactor-specific operating conditions. The unique design requirements of these pumps entail development work on scaled models before design finalisation and prototype manufacture and testing.

This chapter discusses the tests conducted during model development and prototype testing of sodium pumps.

5.2 SIMILARITY

Sodium centrifugal pumps are unique and often require development work. The first step in evolving a new or improved design involves testing on small models. It is essential to realise similarity between the flow conditions in the model and the prototype and this is achieved by enforcing the following similarities. They are:

(i) Geometric similarity – This is achieved by ensuring that the dimensions of corresponding hydraulic parts in the model and prototype are scaled to the same ratio and that all angles are identical in the model and prototype. In addition to ensuring geometric similarity in the pump parts, it is equally vital to ensure geometric similarity in the suction and delivery flow paths. For example, in a vertical pump arrangement, the distance between the pump intake and the near wall is to be simulated so that the intake flow patterns are similar.

(ii) Kinematic similarity – This is achieved by selecting the operating speed of the model so that the velocities at corresponding points in the flow passages in the model and the prototype are scaled to the same ratio.

(iii) Dynamic similarity – This is achieved by ensuring that the forces at corresponding points in the flow passages in the model and the prototype are scaled to the same ratio.

Difficulties may, however, be experienced in specific areas, such as simulating the relative roughness of corresponding parts between the model and prototype or in simulating the clearances between rotating and stationary parts in the model and the prototype. In such cases, it is to be verified that these deviations do not affect the results, of the phenomenon under study, before proceeding further. When dynamic similarity is achieved, it is implicit that both geometric and kinematic similarity exist.

The various dimensionless numbers influencing dynamic similarity between the model and the prototype are Reynolds number, Froude number, and Euler number. The physical process (i.e., performance testing or NPSH testing) under study decides which numbers are to be maintained identical in the model and the prototype.

Reynolds number is the ratio of inertial force to viscous force. In terms of the liquid density, velocity, and characteristic length, it is given by the relation:

$$Re = \rho L^2 U^2 / \mu UL = \rho UL / \mu = UL / \nu$$

where L is the characteristic length, U is the characteristic velocity, ρ is the density of the liquid, and μ, ν are the absolute and kinematic viscosities respectively of the liquid at the operating temperature. For a pump, the characteristic length, L = D, the impeller diameter, and therefore $Re = UD / \nu = ND^2 / \nu$. When the liquid used in model testing is the same as that in the prototype, equality of Reynolds number between the model and the prototype means $(UD)m = (UD)p$. If the model size is small, it becomes difficult to satisfy this condition because the model speed then becomes large and impractical. In such cases care is exercised to ensure that the flow field in the model is turbulent and the order of magnitude of Reynolds number in the model is the same as (or not far from) that in the prototype.

Froude number is the ratio of inertia force to gravity force and is given by the relation:

$$Fr = \rho L^2 U^2 / \rho L^3 g = U^2 / Lg$$

where L is the characteristic length. For a pump, L = H, the pump head. Therefore,

$$Fr = U^2 / gH$$

Euler number is the ratio of inertia force to pressure force. It is given by the relation:

$$\rho L^2 U^2 / \Delta P L^2 = \rho U^2 / \Delta P$$

where ΔP is the pressure change. Euler number equality between the model and the prototype is necessary during cavitation studies.

5.3 DIMENSIONAL ANALYSIS: HEAD AND FLOW COEFFICIENTS

In a centrifugal pump, the variables of interest are the energy per unit mass of liquid pumped, gH; the shaft power, P, and the efficiency, η. These variables are dependent on geometrical parameters of the pump such as impeller outlet diameter, D, a characteristic length, L, surface roughness, ε; operating parameters such as pump speed, N, and pump flow rate, Q; and liquid parameters such as viscosity, μ and density, ρ. The dependent variables can be expressed as:

$$gH = f_1\left(D, L, \varepsilon, N, Q, \mu, \rho\right)$$

$$P = f_2\left(D, L, \varepsilon, N, Q, \mu, \rho\right)$$

$$\eta = f_3\left(D, L, \varepsilon, N, Q, \mu, \rho\right)$$

Using Buckingham's π theorem leads to:

$$gH/N^2D^2 = \phi_1\left(L/D,\ \varepsilon/D, Q/ND^3,\ \rho ND^2/\mu\right)$$

$$P/\rho N^3D^5 = \phi_2\left(L/D, \varepsilon/D, Q/ND^3, \rho ND^2/\mu\right)$$

$$\eta = \phi_1\left(L/D,\ \varepsilon/D,\ Q/ND^3,\ \rho ND^2/\mu\right)$$

In a pump, the flow is highly turbulent, so the effect of the Reynolds number may be neglected it is well above 10000. Similarly, the effect of ε/D may also be neglected because at such high Reynolds numbers ε/D is expected to have the same effect on pumps of commercial size. Moreover, the effect of the irregular shape of the blade passages is more dominant than ε/D; hence, its effect may be neglected. For a geometrically similar pump all dimensions are scaled in the same ratio L/D and therefore this is constant.
    Hence:

$$gH/N^2D^2 = \phi_1\left(Q/ND^3\right) \tag{5.1}$$

$$P/\rho N^3D^5 = \phi_2\left(Q/ND^3\right) \tag{5.2}$$

$$\eta = \phi_3\left(Q/ND^3\right) \tag{5.3}$$

The dimensionless variable gH/N^2D^2 is known as the head coefficient, C_H, while the dimensionless parameter Q/ND^3 is known as C_Q, the flow coefficient. Equations (5.1–5.3) give the relationships between head, flow rate, and efficiency for geometrically similar pumps.

From Equation (5.3) it is seen that at constant efficiency the flow coefficient is constant. From Equation (5.1), in this case the head coefficient is also constant. Eliminating D from the head and flow coefficients gives:

$$\frac{N\sqrt{Q}}{(gH)^{\frac{3}{4}}} = \text{constant}$$

This dimensionless parameter, known as Shape Number, is a function of the pump efficiency. In the pump industry, a more common terminology in use is that of Specific Speed, which is defined as the speed of a geometrically similar pump that delivers unit quantity of liquid at unit head. The Specific Speed, however, is not dimensionless, and it is given by the expression:

$$\frac{N\sqrt{Q}}{(H)^{\frac{3}{4}}}$$

A model pump with dimensions that are a scale factor of the prototype and having all angles between homologous members the same as that in the prototype is dynamically similar to the prototype if its specific speed is the same as that of the prototype.

5.4 AFFINITY LAWS

The affinity laws are used to predict the prototype pump's performance from the test results of the model pump or to predict the operating parameters of a pump operating at any speed from that at the rated speed.

These laws are obtained from equations (5.1, 5.2, 5.3):

$$\left(Q/ND^3\right)_{\text{pump1}} = \left(Q/ND^3\right)_{\text{pump2}}$$

$$\left(gH/N^2D^2\right)_{\text{pump1}} = \left(gH/N^2D^2\right)_{\text{pump2}}$$

$$\eta_{\text{pump1}} = \eta_{\text{pump2}}$$

where 1 and 2 refer to model and prototype, or in the case, of the same pump operating at different speeds, they represent the respective operating rpm.

The prototype performance parameters are extrapolated from the results of model tests using the following relations:

$$Q \propto ND^3$$

$$H \propto N^2 D^2$$

$$P \propto \rho QgH \propto \rho N^3 D^5$$

$$\left(1 - \eta_m\right)/\left(1 - \eta_p\right) = \left(D_p / D_m\right)^{0.26}$$

5.5 FACTORS TO BE CONSIDERED WHILE SELECTING MODEL SCALE AND SPEED

The following are ensured while selecting the size of the model and its speed:

(i) The model is geometrically similar to the prototype pump, meaning that all dimensions are scaled to the same ratio vis-à-vis the prototype, and angles between homologous elements in the model remain the same as in the prototype.

(ii) The model is dynamically similar to the prototype, which is achieved by ensuring the model has the same specific speed as the prototype.

(iii) The nature of flow in the model is similar to that in the prototype. This condition mainly refers to Reynolds number, which in the model shall not be less than one order of magnitude from that in the prototype.

(iv) The model pump head is >80% of the prototype pump head. This condition satisfies requirement (iii) above.

(v) The model has a high surface finish, and the relative roughness(ε/D) of surfaces in the flow passages of the model is not significantly different from that in the prototype (increase in relative roughness not more than 3).

(vi) The rating of the model pump and the geometric scale ratio is in conformance with the pump affinity laws, viz.: $Q \propto ND^3$ and $H \propto N^2 D^2$ where Q-Capacity, H-Head, N-Speed, D-impeller outlet diameter.

(vii) The geometric scale ratio (i.e., size of prototype/size of model) is typically in the range 2–6. Also, the model pump impeller outer diameter is >= 300 mm, in conformity with Hydraulic Institute Standards [1].

5.6 FACTORS TO BE CONSIDERED IN PUMP TESTING

(a) Type of test circuit/loop [2]: Two types of test circuits/loops are used: (i) closed-circuit/loop; and (ii) open circuit/loop.

(i) Open circuit test rig: In its simplest form, the pump draws water from an open sump or reservoir in an open circuit test rig. The water from the pump discharge pipe flows back into the reservoir by gravity over a measuring weir that determines the flow rate. Figure 5.1 [2] shows the sketch of an open circuit test rig.

(ii) Closed-circuit test rig: In its basic form, a closed-circuit test rig consists of a tank with a liquid-free surface, the test pump, and instrumentation to measure the pump parameters. The pumps draw suction from the tank and discharges the liquid back into the tank. Figure 5.2 [2] is a sketch of a basic closed-circuit test rig.

A closed loop has several advantages over an open loop: (i) it can be used for testing of large capacity pumps, which would require huge measuring reservoirs with the open circuit arrangement; (ii) it is safe for pumps handling special/volatile/hazardous liquids (e.g., sodium); (iii) it can be used for testing pumps that operate at high temperatures (e.g., boiler feed pumps); (iv) it permits the lowering of NPSH to values below the atmospheric pressure thus permitting pumps with sub-atmospheric NPSHR to be tested; (v) the air content in the liquid can be controlled to the desired level; (vi) the liquid temperature can be controlled; and (vii) the liquid quality can be maintained. Features (v), (vi), and (vii) are mandatory for detailed cavitation testing such as visual NPSH, acoustic NPSH, and cavitation erosion tests.

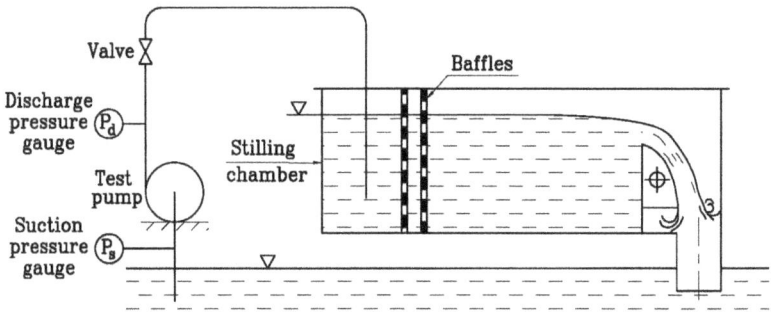

Figure 5.1 Typical open circuit pump test rig.

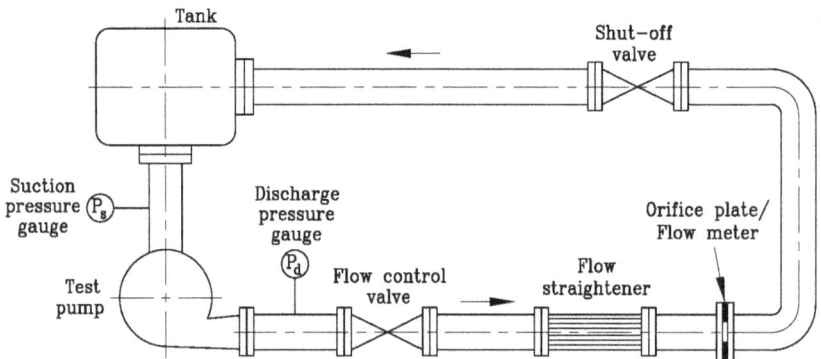

Figure 5.2 Typical closed-circuit pump test rig.

Table 5.1 Comparison of properties of water and sodium [3–5]

Ser	Liquid	Temperature °C (K)	Density kg/m³	Viscosity Pa s× 10³	Kinematic viscosity m²/s²× 10⁶	Vapour pressure Pa
1	Water	20 (293)	998.2	1.002	1.004	2339
		40 (313)	992	0.6526	0.858	7381
		60 (333)	983	0.4666	0.475	19932
2	Sodium	350 (623)	868	0.307	0.353	26.58
		397 (670)	857	0.282	0.329	70.60

(b) Selection of test liquid: Water is the standard test liquid both for model and prototype testing. Since the relevant physical properties, such as density and viscosity, of water at room temperature and that of sodium at the operating temperature are similar (Table 5.1), the pump performance curves obtained with sodium at the operating temperature are identical as that with water at room temperature.

Liquid sodium reacts explosively with moisture, and catches fire on exposure to air. Hence systems that use sodium as the test liquid demand perfect leak tightness, a sensitive leak detection system, and robust sodium firefighting equipment. Furthermore, specialised instrumentation is required for measuring temperature, pressure, and liquid level alongside specific equipment for monitoring and eliminating impurities. Also, the material of construction of the facility is to have good strength and corrosion resistance at high temperatures as well as good weldability. Low carbon austenitic stainless steel (SS304L/LN; SS 316L/LN) is therefore the material of construction in sodium loops. These reasons together make the facility with sodium as the test liquid complicated and expensive.

While water testing is adequate to predict the pump's hydraulic performance, sodium testing helps identify problems related to high-temperature operation. Some such problems include unequal thermal expansion between parts operating with close clearance (such as hydrostatic bearing, wearing ring etc.), distortion due to uneven temperature distribution along the long shaft, the effect of piping reaction on pump verticality, reduction of close clearances in the cover gas space due to sodium vapour deposition, overheating of seals and loss of leak tightness in the cover gas space.

(c) Selection of impeller orientation: Centrifugal sodium pumps are generally of vertical construction, enabling easy sealing of sodium from the atmosphere using a buffer gas space.

Model tests, however, can be done in the horizontal or vertical orientation. It is best to use the model in the same orientation as the prototype because changes in the intake flow path can affect the velocity distribution at the pump inlet, thus affecting pump performance. This problem is especially true for a top suction impeller because here, the velocity of liquid entering the impeller undergoes 180° change in direction, thus adversely affecting the conditions at the impeller eye.

(d) Intake modelling: In cases where multiple pumps draw liquid from the pool that contains other components arranged between the pumps (such as intermediate heat exchangers in the pool-type fast reactor), it is prudent and often necessary to simulate the intake conditions of the pump. These conditions, such as the spacing between the pumps, the submergence of the intake to the pump inlet, the clearance between the pump intake outer diameter and the reactor vessel internal diameter, etc., can affect pump performance due to vortexing, gas entrainment, and other free surface phenomena.

5.7 FEATURES OF A WATER TEST LOOP

A water test loop is used for the determination of (i) pump hydraulic characteristic curves under regular operation; (ii) mechanical characteristics such as pump vibration and unbalanced axial force; (iii) additional pump characteristics in the second quadrant (if operating conditions so demand); (iv) low-flow rate characteristics e.g., minimum operating flow rate to avoid flow surge, recirculation; and (v) conventional cavitation performance, i.e., $NPSHR_{3\%}$, and advanced cavitation testing to determine limits such as cavitation inception, acoustic NPSHR, visual NPSHR, and NPSHR erosion. The water test loop for pump testing will necessarily consist of the following components/systems: (i) a tank with free water level topped by air or a cover gas, upstream of the pump test section; (ii) pump test section; (iii) heat exchanger system for maintaining liquid temperature; (iv) system to control the dissolved air content in the liquid; (v) absorption system to re-absorb the air released in the cavitating zone and entrained in the liquid; (vi) control system for maintaining the flow rate and pressure in the system to desired values; and (vii) instrumentation system for measuring the liquid flow rate, pressure, temperature, and gas content.

In addition to the above features, which are essential for performance measurement, the pump test section is provided with Perspex windows in case visual observation of hydraulic phenomena at the pump suction is also part of the test plan.

Figure 5.3 shows the EPOCA pump water test loop [6] in the EDF laboratory at Chatou in north-central France. This closed-loop facility was used for the testing of both pumps and valves. The pump to be tested is located in the top portion of the loop and draws water from a supply tank located upstream. The flow from the pump passes through the resorber (R1) that re-dissolves the air bubbles released in the cavitation zone. A bank of valves downstream of R1 is used to drop the pump head, and a second resorber (R2) is used to re-dissolve any remaining bubbles before the liquid enters the

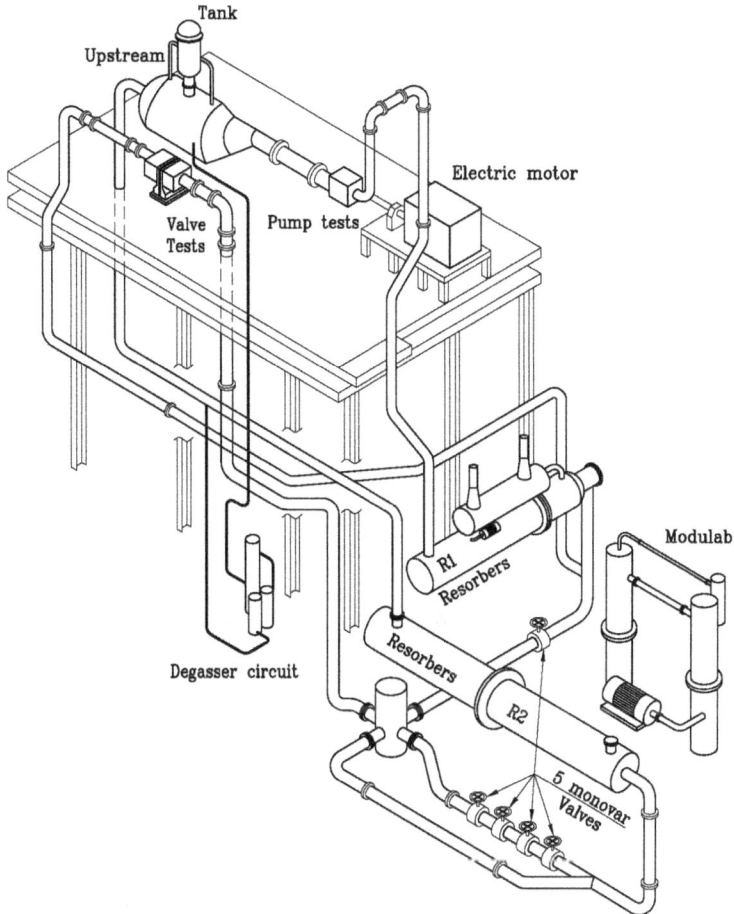

Figure 5.3 EPOCA pump water test loop [6]. (From: Cavitation Bubble Trackers, Lecoffre Y, ©1999, Reproduced by permission of Taylor & Francis Group).

supply tank. The loop is provided with a central control unit that facilitates testing in automatic mode.

5.8 CAVITATION TESTING

The NPSHA in a sodium system is limited because:

(a) sodium systems are low-pressure systems;
(b) pump submergence is limited, and
(c) cover gas pressure is only marginally above atmospheric pressure to facilitate easy sealing of radioactive cover gas from the atmosphere.

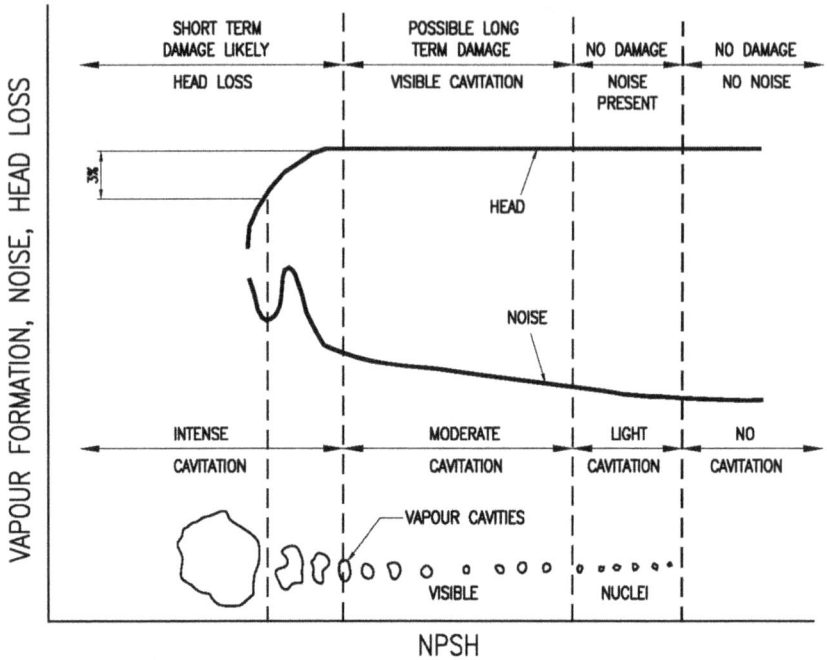

Figure 5.4 Effect of NPSH reduction on pump head, cavitation noise, and bubble size [7] (Credit: Courtesy of the Nuclear Institute – Sketches from Liquid Metal Engineering Technology conferences organised by The British Nuclear Energy Society (BNES)).

Since sodium pumps must operate in this modest NPSHA environment for long periods without maintenance, it is important to measure not only conventional NPSHR $_{3\%}$ (3% head drop) performance but to also measure other NPSH parameters such as $NPSHR_{acoustic}$, $NPSHR_{visual}$, $NPSHR_{erosion}$ and $NPSHR_{0\%}$ in addition to $NPSHR_{3\%}$. Figure 5.4 [7] shows the effect of reduction in NPSH on the pump head, cavitation noise, and bubble size.

5.8.1 Test rig for cavitation testing

Figure 5.5 shows the sketch of a typical test rig for cavitation testing in water.

A test rig for cavitation testing is to necessarily have the following provisions:

(i) temperature control; (ii) dissolved gas content control; and (iii) provision for varying NPSH at the pump suction. In case visual NPSH tests are to be conducted, the suction passages of the rig and the pump are to be

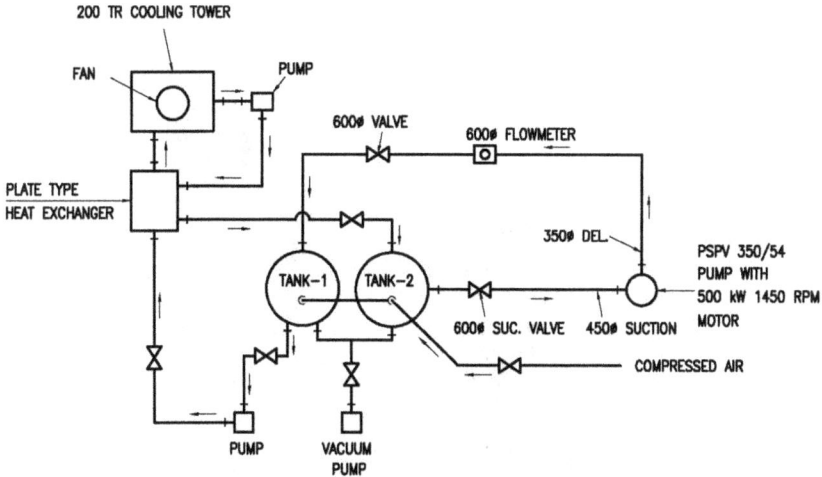

Figure 5.5 Cavitation test facility [8]. (Reproduced from S.G Joshi, A.S Pujari, R.D Kale and B.K Sreedhar, Cavitation studies on a model of primary sodium pump, Proceedings of FEDSM02, The 2002 joint US ASME European Fluids Engineering Summer Conference, July 14–18, 2002, Montreal, Canada).

provided with Perspex observation ports for measurement of the cavity length.

Temperature control is essential in cavitation testing; water tests are usually conducted below 30°C. A low temperature is preferred because the measured NPSHR$_{3\%}$ tends to be low (less conservative) with increased liquid temperature. The reason is that as the temperature of the liquid increases, the ratio of the specific volume of vapour to the specific volume of the liquid decreases, and therefore the effect of insufficient NPSH on pump performance, as manifested by a drop in head, becomes smaller.

In the case of NPSH$_{visual}$ tests, the impeller is provided with a transparent front shroud. Likewise, the pump intake/suction pipe is provided with Perspex windows to facilitate observation/measurement of cavity dimensions as the NPSH is varied. Figure 5.6 shows the transparent portions on the intake skirt, suction bell, and pump impeller (front shroud), facilitating visual observation of pump suction.

In experiments for cavity length measurement, a grid of known dimensions is painted on the impeller surface. The pump is operated under cavitating conditions, and the impeller suction face is viewed under stroboscopic lighting. The dimensions of the cavity patch are measured using the grid as a reference. Empirical relations in literature [9] are then used to estimate the impeller life based on the measured cavity dimensions. Figure 5.7 shows the grid plotted on the blade suction face.

Figure 5.6 Transparent windows on pump intake skirt to facilitate observation of impeller [8]. (Reproduced from S.G Joshi, A.S Pujari, R.D Kale, and B.K Sreedhar, Cavitation studies on a model of primary sodium pump, Proceedings of FEDSM02, The 2002 joint US ASME European Fluids Engineering Summer Conference, July 14–18, 2002, Montreal, Canada).

5.9 TYPES OF TESTS

The tests done on a reactor pump (model and prototype) include:

(a) Performance testing: Establishing the pump head vs flow rate, pumping power vs flow rate, and pump efficiency vs flow rate characteristics. Model sizing, testing procedure/instrumentation, and acceptance criteria are to conform with the code (HIS, DIN, ISO, BIS, to name a few) agreed upon by the client and pump manufacturer.

   The extrapolated performance of the prototype pump is confirmed during prototype pump testing.

(b) Cavitation testing: Establishing the $NPSHR_{3\%}$ vs flow rate characteristics. The procedure for performance testing and conventional

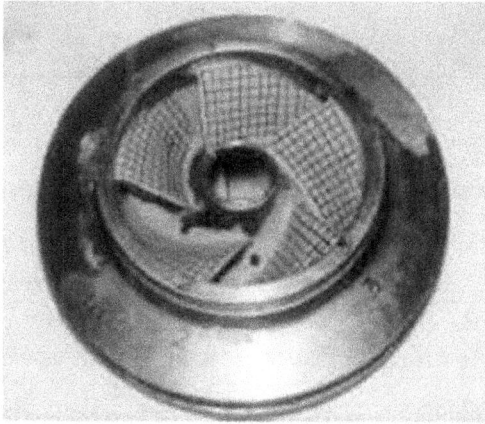

Figure 5.7 Grid on suction face for cavity length measurement [8]. (Reproduced from S.G Joshi, A.S Pujari, R.D Kale and B.K Sreedhar, Cavitation studies on a model of primary sodium pump, Proceedings of FEDSM02, The 2002 joint US ASME European Fluids Engineering Summer Conference, July 14–18, 2002, Montreal, Canada).

cavitation testing (NPSHR$_{3\%t}$) are specified in standards (e.g., Hydraulic Institute Standards (HIS) [1], ISO, DIN, API). Additional tests are also done to establish the NPSH at which: (i) acoustic noise (NPSH$_{acoustic}$) occurs, signifying the inception of cavitation; (ii) bubbles are detected (NPSH$_{visual}$); and (iii) erosion due to cavitation is maximum (NPSH$_{erosion}$). Long-term erosion tests using scaled models are necessary to establish the variation of erosion rate of the impeller with NPSH. This difficulty is overcome by coating the impeller with special paint that will erode only under cavitating conditions and using paint removal as a conservative indicator of the onset of erosion. Paint erosion tests are most useful in identifying cavitation erosion susceptible areas on the pump impeller.

While the above tests can be performed on the model pump, the testing of the prototype pump is generally limited to conventional NPSHR$_{3\%}$ testing.

(c) Measurement of residual axial thrust: In the case of a bottom suction impeller, the axial thrust during operation can combine with the weight of the rotating parts to produce a formidable thrust in the downward direction. In the case of a top suction impeller, however, the axial thrust will oppose the self-weight of the rotating assembly during operation resulting in a possible change in the direction of axial thrust at an intermediate operating speed. A design option to reduce the axial thrust is to provide balancing holes in the impeller. Axial thrust measurements during performance testing are useful in confirming

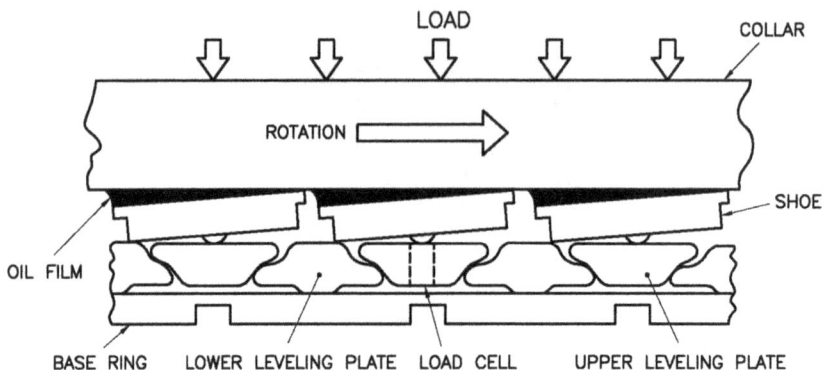

Figure 5.8 Load cell for measuring axial thrust in Kingsbury bearing.

the adequacy of the capacity of the thrust bearings. Kingsbury tilting pad type thrust bearing, commonly used in large pumps, is shown in Figure 5.8. In this arrangement, the thrust load is measured by installing a load cell in the levelling plate (Figure 5.8) of the prototype pump. The load cell and levelling plate are calibrated as a unit to maximise accuracy. Since the thrust load on the bearing is uniformly distributed among the shoes, the rating of the load cell used is only a fraction of the load capacity of the bearing. For instance, in a six-shoe bearing the capacity of the load cell is only ⅙ of the total load.

(d) Endurance testing at rated flow rate: This test on the prototype pump is done at the rated flow and speed at the plant NPSH for an extended duration, decided by the designer, and finalised after mutual discussions between the user and the supplier. The endurance test duration depends on the power rating of the pump, and a typical endurance test is of 100 h–250 h duration.

(e) Endurance testing at maximum flow rate: This test is done on the prototype pump at the rated speed at the maximum flow rate expected in the reactor and the corresponding NPSH margin for 24 h–48 h duration.

(f) Pump coastdown test: This test is done on the prototype pump at the rated speed and flow rate. After stable flow conditions are achieved, significant parameters such as pump head, water level in pump tank, cover gas pressure, liquid temperature, and bearing parameters are recorded. The motor is then tripped, and the variation of pump speed vs time, pump flow rate vs time are recorded.

(g) Strip test: After all tests are completed, the pump is dismantled and subjected to a thorough visual examination. A more detailed examination is done in the event of observation of scoring marks/damage to parts.

(h) Testing of pump to pipe connection: The primary pump in a pool-type reactor is connected to the reactor grid piping through a non-permanent joint. The connection is designed to facilitate easy assembly and removal of the pump from the reactor (please refer to Figures 3.11–3.15 in Section 3.2.5.2) and ensure minimal leakage of high-pressure sodium to the cold pool without cavitation. Testing the joint during development is necessary to finalise the hydraulic geometry, measure the leakage flow rate through the joint, and confirm cavitation-free operation (through vibration and noise measurements). Additional tests to evaluate mechanical performance may be required depending on the type of joint selected.

(i) Testing of pump support: In pool-type reactors, the primary pumps are suspended from the roof slab of the reactor with the pump discharge coupled to the reactor inlet piping (please see Section 3.2.5.1 for design options to support primary pumps in pool-type reactors). The supporting arrangement should accommodate the differential thermal expansion without stressing the support and its interface with the roof slab. The mechanical testing of the support is done by testing the pump in sodium at the operating temperature (or in water at room temperature) while simulating the pump/system interface in the reactor.

Tests mentioned in (h) and (i) above are additional tests for validation of specific design concepts employed in centrifugal sodium pumps and are not part of general pump performance testing.

The full-scale rotating assembly of the SuperPhénix-1 (SPX-1) primary pump was done in the PIVOTERIE-1 test setup [10]. The test arrangement consisted of a vertically mounted test section containing the pump rotating assembly, including a disc (dummy impeller) of the same inertia as the pump impeller, in-sodium hydrostatic bearing, above sodium roller bearings and seals, and flexible coupling. The test setup ensured both geometric and thermal similitude with the actual pump. Independent pumps located outside the test vessel provided the high-pressure feed to the hydrostatic bearing. The test setup was provided with extensive instrumentation, including thermocouples for temperature measurement at various elevations, accelerometers for vibration measurement, and specially designed inductive probes for in-sodium measurement of shaft instantaneous position within the hydrostatic bearing. The test section was supported on pivots to facilitate tilting of the pump rotating assembly and thereby simulate the inclined operation of the pump in the reactor.

The tests conducted included the study of: (i) shaft behaviour under simulated reactor conditions vis-à-vis pressure, temperature, and speed of operation; (ii) shaft over speed behaviour; (iii) behaviour of hydrostatic bearing, top bearing, and seals; (iv) inclined operation of rotating assembly; (v) behaviour of rotating assembly under seismic loading; (vi) assembly and disassembly procedures; and (vii) procedure for sodium removal from the rotating assembly. Figure 5.9 shows the PIVOTERIE-1 test section.

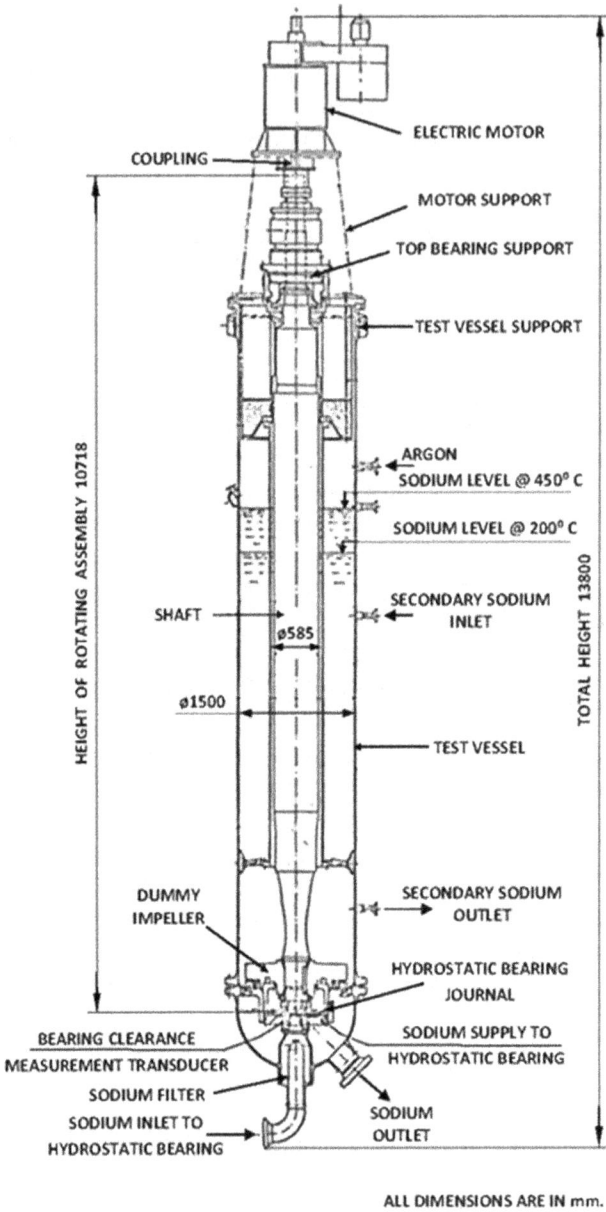

ALL DIMENSIONS ARE IN mm.

Figure 5.9 PIVOTERIE-1 test section [10] (Credit: Courtesy of the Nuclear Institute – Sketches from Liquid Metal Engineering Technology conferences organised by The British Nuclear Energy Society (BNES)).

5.10 MODEL TESTING WITH AIR AS THE WORKING FLUID

The use of air, instead of water, as the working fluid during model tests has the following advantages:

  (i) Lower cost of models (as the models can be shaped easily from materials such as wood, plastic, etc.,) and test rig.
 (ii) Lower energy consumption.

The disadvantages are:

  (i) Since air density is about 800 times less than that of water, the head generated with air as the working fluid is small and prone to measurement error.
 (ii) Because of the significant difference in the Reynolds number between water and air, the accuracy of the calculated value of prototype efficiency is poor.
(iii) The relative mechanical losses will be disproportionately high because the power delivered to the pump with air is only 1/800 that with water for the same flow rate and head.

The above disadvantages are overcome by using compressed air as the working fluid. The increased density of compressed air results in more accurate head measurement and reduced relative mechanical losses. As the air pressure increases, the kinematic viscosity decreases, and the value of Reynolds number increases to a value close to that with water as the working fluid.

However, if the pump has a bearing that depends on the fluid pressure for bearing action (e.g., hydrostatic bearing), then the load capacity of the bearing could possibly be insufficient, and it may be required to replace it with a conventional bearing. Lastly, cavitation testing of the pump is not possible with air as the medium.

Air tests on 3/16 scale model were done during the hydraulic development of the 85,000-gpm sodium pump. The tests were useful in determining the H-Q characteristics of the pump, radial thrust, and locked rotor impedance [7].

Model testing using air as the working fluid was used in scaled-down tests to develop BOR-60, BN-350, BN-600 and BN-800 primary and secondary pumps [11].

5.11 NEED FOR SODIUM TESTING OF PUMPS

As mentioned in Section 5.6(b), the performance curves with sodium (H/Q, P/Q, η/Q, $NPSHR_{3\%}$/Q) are the same as with water. There can even be a marginal increase in efficiency with sodium because of the effect of temperature on hydraulic losses. However, sodium testing is invaluable in understanding the effect of temperature on the mechanical operation of the pump. These include:

(i) The effect of uneven expansion under temperature on the functioning of parts moving relative to each other and operating with close clearances, such as hydrostatic bearing, wearing rings, etc.

(ii) The effect of convection in the cover gas region of the pump that can produce temperature asymmetry in the circumferential direction and possible bowing of the pump shaft.

(iii) Thermal shocks (hot/cold) resulting from conditions such as plant start-up/shutdown, reactor scram, components malfunction, one pump trip, etc. which can cause mechanical problems due to reduction in close clearances because of the unequal expansion of relatively moving parts with different thermal inertia from the sudden temperature change.

However, setting up a dedicated sodium test facility for pumps is capital-intensive. Some questions which can assist in arriving at a sound decision are:

(i) Can the facility be combined with that for testing other components?

(ii) Is a full-scale facility required, or will an intermediate-size facility provide the necessary information at a lower cost? (For example, the USSR adopted an intermediate-size facility for testing pumps of the BN reactors.)

(iii) What are the long-term objectives of the reactor programme, and can the facility serve future needs?

(iv) How much experience from similar facilities built elsewhere can be utilised in the development of a new sodium pump design?

In the Indian FBR program, following careful deliberation, it was decided to select an alternative approach instead of full-scale sodium testing of the prototype pumps. This approach focused on the following aspects: (a) design, fabrication, and testing of a small centrifugal sodium pump in both water and sodium, (b) testing of model and prototype hydraulics in water, (c) testing of the rotodynamic behaviour of a representative rotating assembly, with dummy impeller, mounted on a spherical seat to validate the concept, and (d) fabrication, including heat treatment, of critical components such as impeller, diffuser, shaft and hydrostatic bearing. Additionally, valuable insights were gained from the simultaneous operation and troubleshooting of the primary and secondary centrifugal pumps at the Fast Breeder Test Reactor.

5.12 FEATURES OF A LOOP FOR SODIUM TESTING OF PUMPS

A sodium loop, compared to a water loop, has several unique features to exploit the special properties of liquid sodium, such as its metallic nature, higher than room temperature melting point, good electrical and thermal conductivity, as well as effective protection from its high chemical activity with air/water. These features are:

1. Additional tanks to dump the sodium and empty the loop after testing.
2. Argon cover gas system to blanket the sodium-free surface in the tanks and prevent air ingress.
3. Instruments (eg., plugging indicator) to monitor the impurity level in sodium during operation.
4. Equipment (eg., cold trap) to remove impurities from sodium and maintain the desired level of sodium purity during operation.
5. Specialised instrumentation, that exploit the electrical and magnetic properties of the liquid metal, to monitor sodium levels in the tanks and measure loop parameters such as flow rate, pressure, and temperature.
6. An electrical system dedicated to heating as well as maintaining the sodium in the loop in the liquid state at the desired temperature.
7. Auxiliary pumps, typically of electromagnetic type, for preliminary filling of the loop.
8. Specialised instrumentation for the detection of any leakage of sodium into the atmosphere (e.g., sodium aerosol detector).
9. A dedicated external system for sodium firefighting. A three-stage approach of prevention, detection, and mitigation is adopted; moreover specific powder extinguishers are used for extinguishing sodium fires such as Sodium bicarbonate or Dry Chemical Powder, Ternary Eutectic Chloride (TEC), Sodium chloride, and graphite-based powders.

Figure 5.10 is a schematic of a typical sodium pump testing facility.

The facility consists of a storage/dump tank, an additional feed tank, a pump tank, a purification circuit, and a cover gas system.

The storage/dump tank is located at the lowest elevation, and its function is to contain the entire quantity of sodium in the loop when the loop is drained for maintenance/repair. An additional feed tank of smaller capacity is provided to supply sodium at a temperature different from the sodium in the main loop to simulate cold/hot thermal shock in the test pump. The feed tank elevation, vis-à-vis the test pump suction line, is located such that the sodium feed rate to generate thermal shock at the pump suction is achieved by regulating the argon gas pressure in the feed tank. The pump tank houses the test pump, simulating the pump-free surface level, the intake geometry, and pump suction conditions. The purification circuit consists of a cold trap, economiser, plugging indicator, and sodium sampler. A fraction of the sodium from the main loop is cooled in the economiser before entering the cold trap, where oxide and hydride impurities are precipitated in a wire mesh. The sodium emerging from the cold trap is re-heated in the economiser before it re-enters the main loop. A plugging indicator measures the impurity level in the sodium by monitoring the temperature at which the impurity precipitates. The cover gas system maintains the blanket of cover gas over the sodium-free surface in the various tanks and prevents air ingress

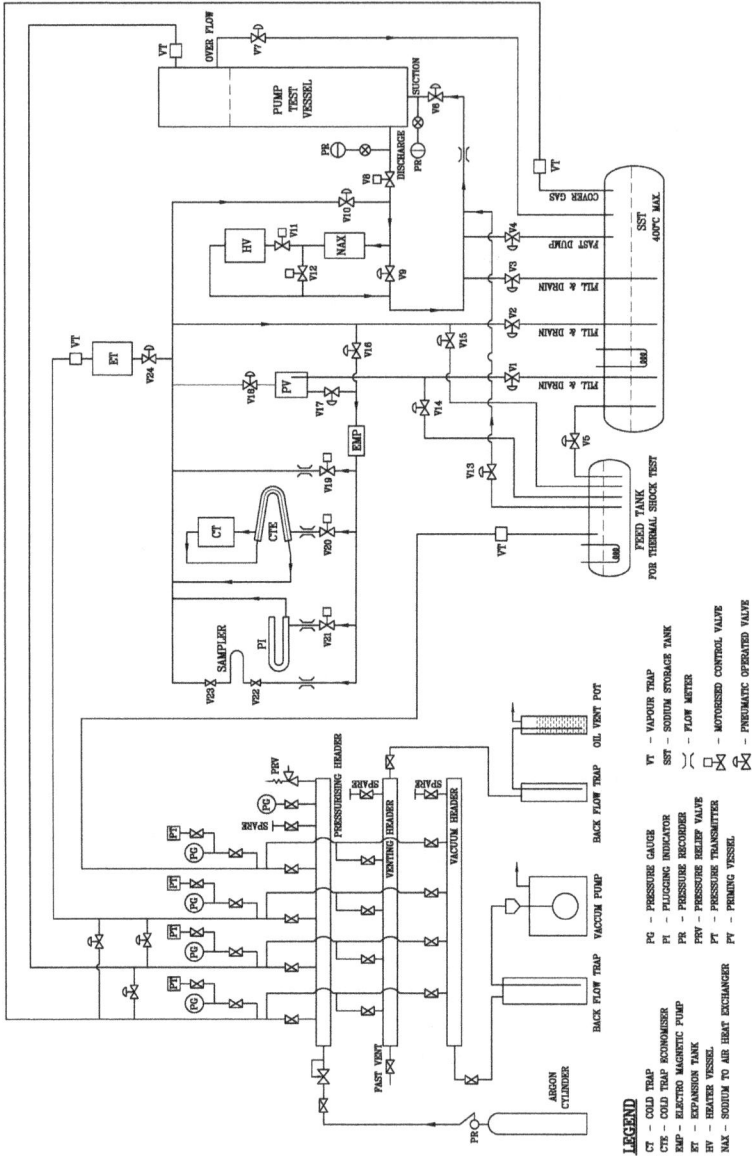

Figure 5.10 Typical facility for testing of pumps in sodium.

into the system. An electromagnetic pump is used to fill the sodium in the loop and maintain circulation in the purification system. A priming vessel is provided at the suction of the EM pump to prime the pump prior to starting. An expansion tank is located at the topmost point of the loop to accommodate the volume change in sodium with temperature. Vapour traps are provided in the cover gas lines at the outlet of each vessel to entrap sodium vapour and prevent clogging of the gas lines. The entire system is provided with surface heaters to preheat the system and insulated to reduce heat loss to the environment. Immersion heaters are provided in the dump tank and the feed tank to raise the temperature of sodium quickly to the required value. Leak detectors are provided at critical locations (such as weld joints and high-stress regions) to monitor for sodium leaks. A dedicated heater vessel and sodium-to-air heat exchanger are provided in the test pump circuit to regulate the temperature of sodium in the pump test loop.

5.13 PUMP SODIUM TESTING FACILITIES WORLDWIDE

This section discusses some of the facilities installed worldwide, during the height of the fast reactor programme, for testing large centrifugal pumps in sodium.

(i) Sodium pump Test facility (SPTF) at Santa Susanna, California, USA [12–14]:

This facility, established by Liquid Metal Engineering Center (LMEC), later known as Energy Technology Engineering Center (ETEC), was set up to serve as a test bed for the development, testing, and performance verification of full-scale, liquid metal coolant pumps. The SPTF complex comprised of:

(a) The pump test loop.
(b) The component handling and cleaning facility
(c) A dedicated electrical sub-station.

The pump testing loop was designed for proof/acceptance testing of prototype pumps with a flow rate of up to 120,000 gpm. (7.56 m3/s) and temperature up to 1200°F (648.9°C).

The loop was designed to determine:

(a) Pump characteristics.
(b) Speed-flow rate control response.
(c) Pump performance under thermal shock.
(d) Performance under extended operational tests.

In addition to full-scale pump testing, the facility was also suited for testing of large valves, instruments, and other components. The facility was designed for testing of two pumps, either independently or simultaneously, with flow rates ranging from 20,000 gpm (1.26 m3/s) to 60,000 gpm (3.78 m3/s) and with modifications for testing of a single pump of flow rate 120,000 gpm (7.56 m3/s).

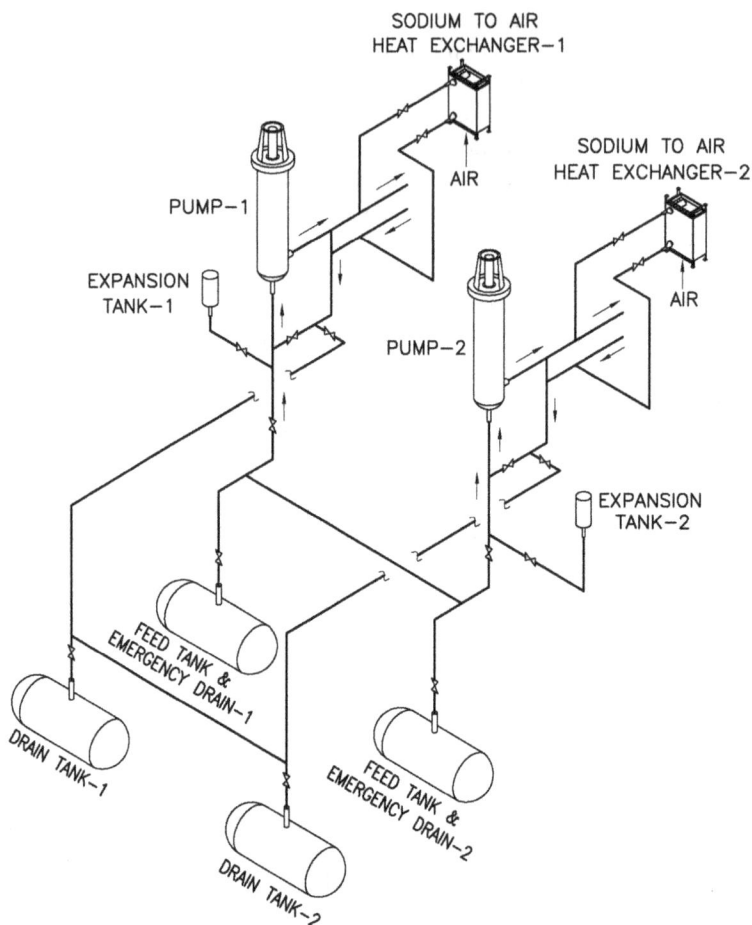

Figure 5.11 Sketch of SPTF.

The facility consists of two loops, as shown in Figure 5.11. The loop is 30" (762 mm) in diameter and consists of a drain tank, feed tank, pump test vessel (in which the test pump is mounted), expansion tank, heaters, coolers, and purification circuit. The drain tank is used to dump the loop after completion of testing, while the feed tank is used to generate thermal transient.

The expansion tanks accommodate the change in sodium volume with temperature as well as the changes in sodium level in the pump tank and permit each loop to operate independently. The expansion tank is also employed during regular operation to control the pressure at pump suction by regulating the pressure of the argon cover gas.

The loop was designed to simulate thermal transients experienced during normal plant startup and shutdown, reactor scram, and conditions resulting from malfunction or failure of components (such as pump, valve, heat exchanger, etc.). Each loop had a feed tank, a drain tank, and the required flow control valves. Hot/cold thermal transient in the pump was simulated by feeding sodium from the feed tank (at a temperature different from that in the main loop) into the pump suction piping by pressurising the feed tank and simultaneously withdrawing the excess sodium from the loop into a drain tank. The pressures in the feed tanks were controlled during these tests to minimise the variation in the pump suction pressure. The drain tanks also enabled quick draining of the main loops during an emergency. An integrated sodium purification system employing filters, hot traps, and cold traps was provided to maintain the required sodium purity during operation.

The pump test facility also had an integrated sodium removal facility. After completion of tests, the pump was transferred to the cleaning tank in the sodium removal facility using the overhead cranes. The sodium removal facility used the alcohol cleaning process to remove sodium from the wetted pumps.

This facility was used to test the prototype pumps of the following reactors, viz. FFTF [15], CRBRP main pump [16], 85,000 gpm inducer pump for large LMFBR [17], etc.

(ii) APB pump test facility at Bensberg, Germany [18, 19]: This facility (Figure 5.12) was constructed by Interatom for sodium testing of pumps of SNR-300 reactors with a flow rate of 5000 m³/h and was later augmented to test pumps with capacity up to 20,000 m³/h. The facility consisted of a main loop of 600 mm diameter and a parallel loop of 350 mm diameter with a pump tank, expansion vessel, thermal shock vessel, purification circuit, and drain tank. The parallel path of 350 mm diameter was used for low-flow rate application. The pump pressure was dissipated using two valves in series (IV1 and IV2) with a fixed-pressure drop device. Suction and discharge pressure was measured using a calibrated transmitter separated from the sodium-filled Bourdon tube by a Na–K-filled chamber. The test pump flow rate was measured using a venturi flow meter, and the sodium level in the pump tank using a continuous level indicator. The other instrumentation provided facilitated the measurement of parameters such as pump vibrations, pump shaft movement, pump speed, pump torque, pump cavitation noise, and sodium temperature at multiple locations in the loop. The liquid leaking past the hydrostatic bearing overflows from the pump vessel into the expansion vessel, which is connected to the main circuit. The NPSH available at the pump suction is the sum total of the cover gas pressure in the expansion vessel, the height of the sodium column in the expansion vessel above the pump suction line minus the losses in the return line from the expansion vessel to the pump suction. Variation in NPSHA during the cavitation test was done by throttling valve IV4 in the return line from the expansion vessel. The purity of sodium in the loop was achieved using a cold trap. A separate vessel was provided to maintain sodium at a temperature

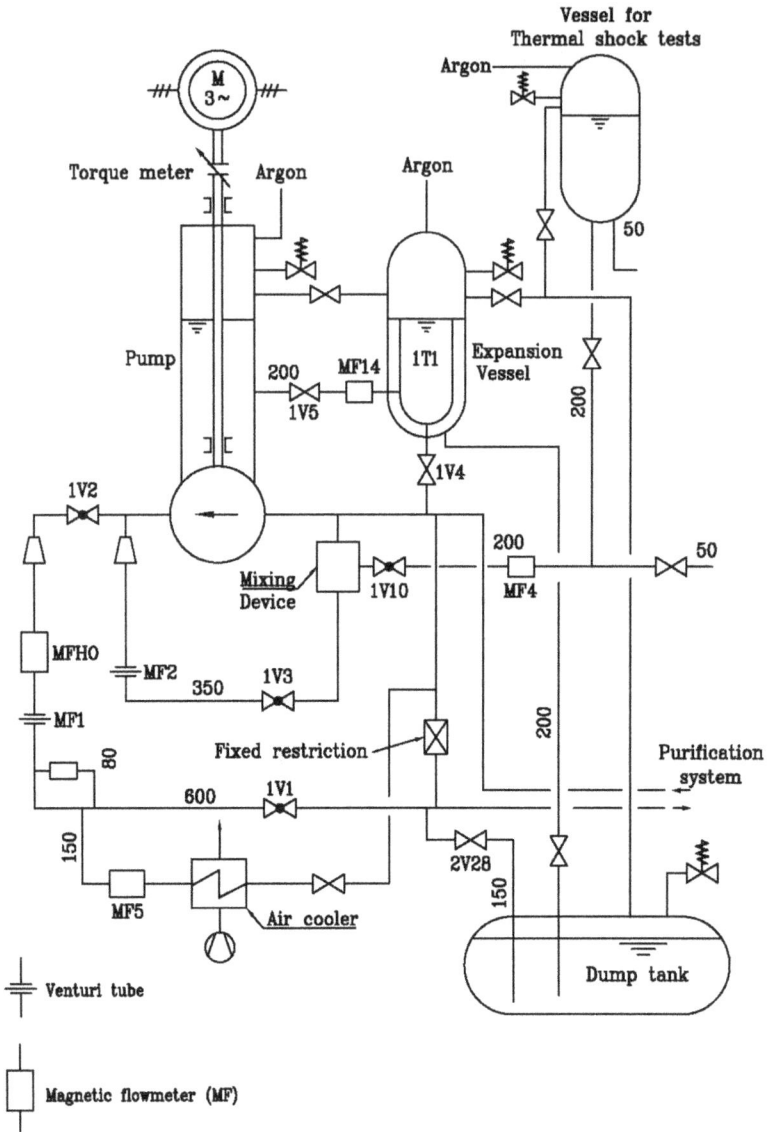

Figure 5.12 APB sodium pump test facility at Bensberg [19]. (Re-published with permission from the proceedings of Pumps for Nuclear Power Plants, IMechE Convention, University of Bath, 22–24 April, 1974).

different from that of sodium in the main loop. Thermal shock was simulated by injecting the sodium from this vessel into the pump suction through a mixing tee. The material of construction for the pump tank, piping, and components was austenitic steel SS304, while the dump tank was fabricated from Ferritic steel (15Mo3).

(iii) Pump testing rig at Reactor Engineering Laboratory (REL) at Risley, Warrington, United Kingdom [14, 20]:

In this facility, the vertical pump under testing was enclosed in a pump tank of diameter 200 cm (Figure 5.13). The tank simulated the flow conditions at the pump inlet, with a free sodium surface in the pump, and the pump supporting arrangement. The main loop was 12 inches (300 mm) in diameter and constructed from stabilised stainless steel 18/8/Ti, and designed to operate at 400°C. The pump tank and the 300 mm pipe were enveloped with a mild steel duct/jacket. Air circulation through the jacket facilitated the heating and cooling of the system; the double enclosure also served as a secondary envelope in the case of a sodium leak. Two storage tanks located in mild steel-lined concrete pits were provided to contain the total quantity of sodium in the system (~12 tons) after dumping of the loop.

The flow rate was monitored using a venturi flow meter and an electromagnetic flow meter that was later replaced by a saddle coil

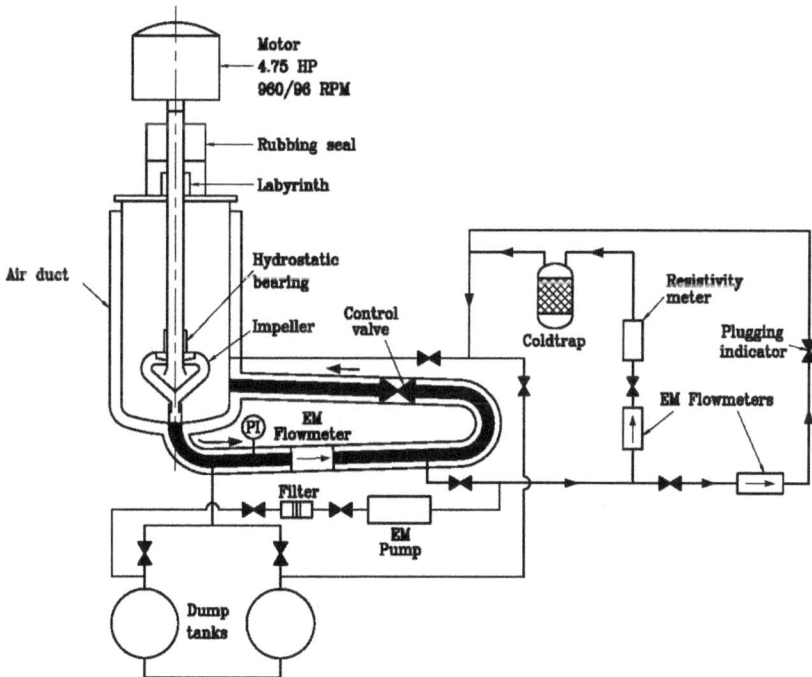

Figure 5.13 Sodium pump test facility at Reactor Engineering laboratory (REL), UK [20]. (Re-published with permission from the proceedings of Pumps for Nuclear Power Plants, IMechE Convention, University of Bath, 22–24 April, 1974).

flowmeter. Capacity regulation was facilitated by a flow control valve that was later replaced with a fixed-pressure drop device.

The purity of the sodium was maintained using a cold trap in an auxiliary circuit through which sodium was circulated using an electromagnetic pump. Operational purity was monitored using a plugging indicator. Oxygen purity was monitored separately using a resistivity meter; impurity levels were also monitored by sampling and laboratory analysis.

(iv) Pump test facilities in Japan [21]: Sodium test facilities of various sizes were constructed in Japan towards the development of sodium pump technology. The first facility to be erected was the small sodium pump test loop at the Hitachi Research laboratory. This loop of size 100 mm NB was used to study the behaviour of a small capacity pump (Q=1m³/min, H = 40 m, N = 2600 rpm). The test pump is installed in a vertical tank with an external heater vessel for raising the temperature of sodium (Figure 5.14). The pump takes suction from the tank (sump type) and discharges into the loop piping; the liquid flows back into the tank to form a closed-circuit. No specific level control is required

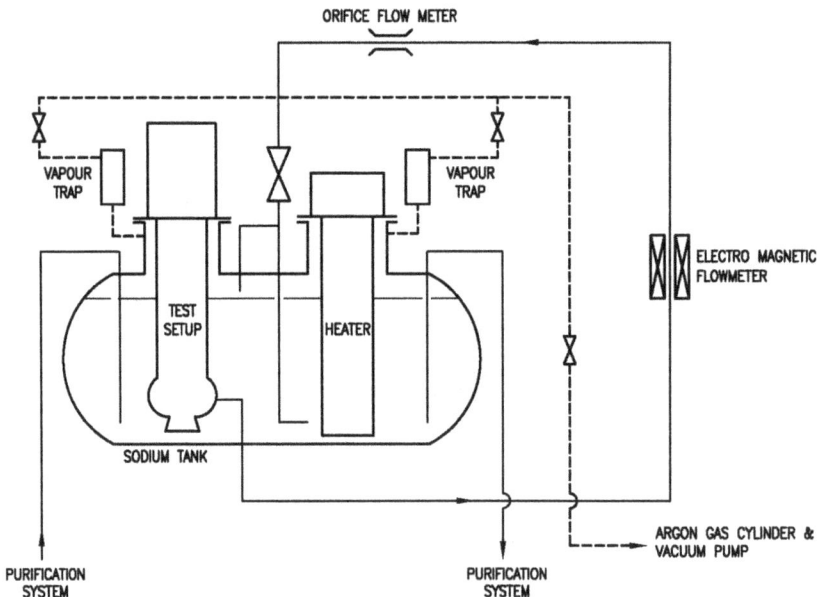

Figure 5.14 Small pump test loop [21] (Oda, Y., Suzuki, A., Kawahara, S., Sagawa, N., Komatsu, T., Ito, Y., "Development of mechanical sodium pump", Sodium-Cooled Fast Reactor Engineering, Proceedings of a Symposium Held in Monaco, 23–27 March 1970, IAEA, Vienna (1970) 375–396. Reproduced with kind permission of the International Atomic Energy Agency (IAEA)).

because the sodium pumped from the tank returns to the tank. The pump flow rate is measured using orifice and electromagnetic flow meters; the pump discharge pressure is measured using a Na–K filled pressure gauge; the pump speed is measured using a tachometer and the pump input power is measured using a strain gauge type torque transducer. Bellows is provided between the pump discharge piping and pump tank to accommodate differential thermal expansion.

The Sodium Test Loop (STL) established at the Oharai laboratory of Power Reactor and Nuclear Fuel Development Corporation (PNC) was erected for testing large-sized pumps. The loop schematic is shown in Figure 5.15. An electromagnetic pump is used to fill the loop and prime the test pump. The loop of size 250 mm NB is provided with a dedicated purification circuit. The facility was used for testing scaled pumps (¼ scale) of the Japanese Experimental Breeder Reactor (JEFR) and for operator training.

A large sodium pump test loop (SPTL) (Figure 5.16) was also established at the Oharai Laboratory by PNC for testing of the large pumps of the Japanese Prototype Fast Reactor (JPFR) and the main coolant pumps of the JOYO reactor.

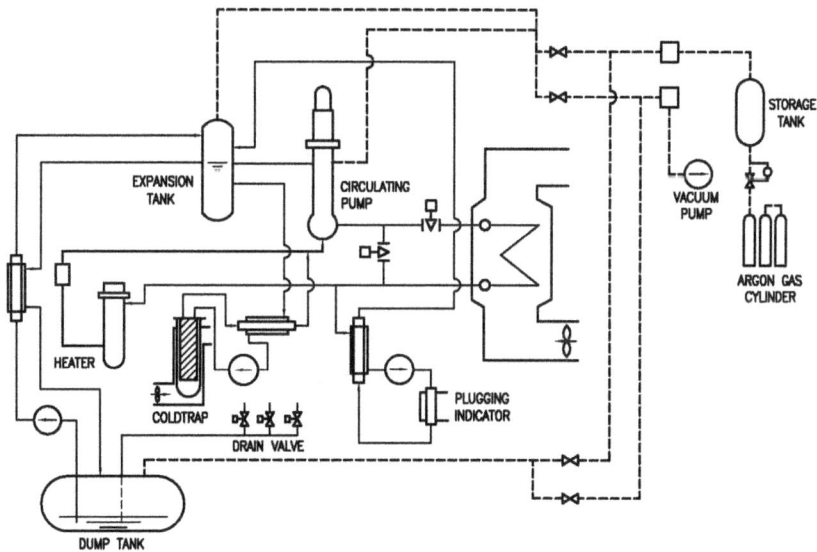

Figure 5.15 Sodium Test Loop (STL) [21]. (Oda, Y., Suzuki, A., Kawahara, S., Sagawa, N., Komatsu, T., Ito, Y., "Development of mechanical sodium pump", Sodium-Cooled Fast Reactor Engineering, Proceedings of a Symposium Held in Monaco, 23–27 March 1970, IAEA, Vienna (1970) 375–396. Reproduced with kind permission of the International Atomic Energy Agency (IAEA)).

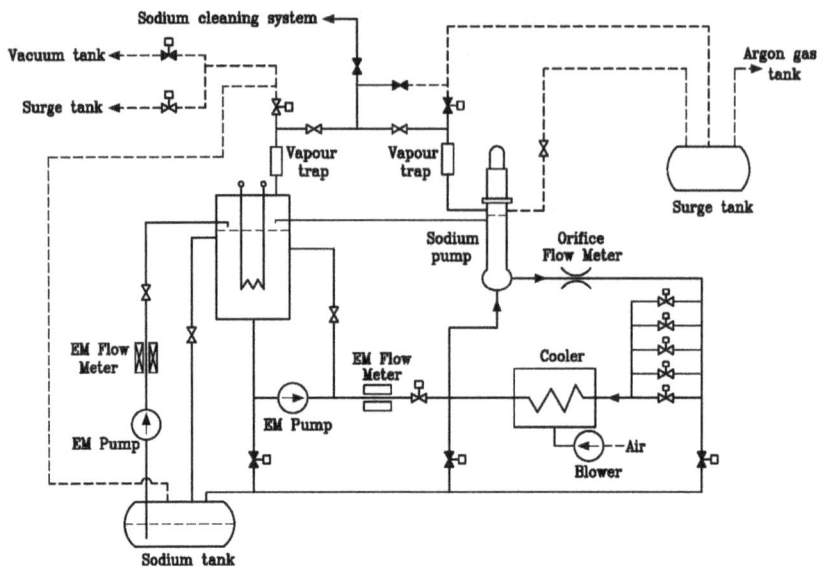

Figure 5.16 Flow sheet of Large Sodium Pump Test Loop (SPTL) [21]. (Oda, Y., Suzuki, A., Kawahara, S., Sagawa, N., Komatsu, T., Ito, Y., "Development of mechanical sodium pump", Sodium-Cooled Fast Reactor Engineering, Proceedings of a Symposium Held in Monaco, 23–27 March 1970, IAEA, Vienna (1970) 375–396. Reproduced with kind permission of the International Atomic Energy Agency (IAEA)).

The loop piping was of size 300 mm NB and the major ratings of the loop were design flow rate = 22 m³/min, design temperature = 550°C, design pressure = 10 kg/cm²

It was designed for the following tests:
- Steady-state performance testing at 440°C.
- Variable flow rate and variable pressure testing.
- Transient testing.
- Start and stop testing.
- Endurance (continuous operation) test.

(v) Facilities for testing sodium pumps of SuperPhénix reactors of France [22–24]:

Sodium testing of the full-scale rotor assembly of the SuperPhénix-1 primary pump was done at the CPV-1 rig built by FIAT-TTG at Brasimone Energy Research Centre. The full-scale rotating assembly (refer Figure 5.9), complete with hydrostatic bearing and seals and a dummy impeller (of equal mass and moment of inertia as the actual impeller), was tested in sodium while simulating reactor operating conditions such as temperature, sodium level, shaft inclination etc. to understand the behaviour of the pump rotor under normal and accident conditions. The tests provided feedback on the performance of

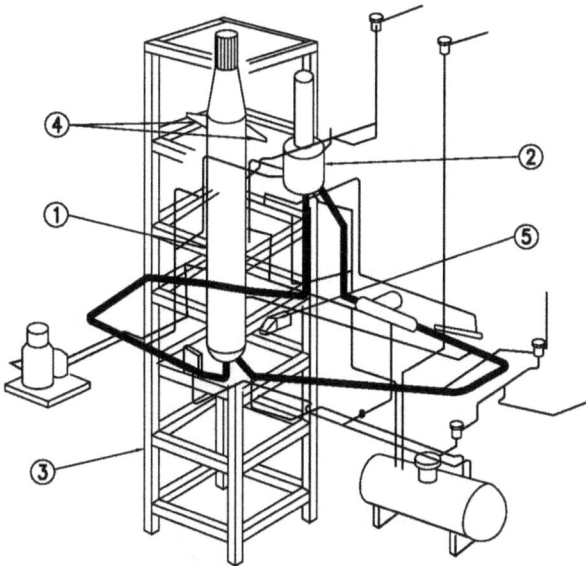

1. Pivoterie–I vessel
2. Duty pump
3. Metal frame
4. Hinge pins
5. Electrohydraulic actuator

Figure 5.17 Full-scale rotor assembly test rig [22]. (Credit: Courtesy of the Nuclear Institute – Sketches from Liquid Metal Engineering Technology conferences organised by The British Nuclear Energy Society (BNES)).

hydrostatic bearing with repeated starts/stops, endurance testing, and rotor dynamics. The test rig is shown in Figure 5.17.

One of the secondary circuits in the RAPSODIE reactor was used as a test bed for cavitation testing of the SuperPhénix-2 (SPX-2) pump impeller. Realising the importance of sodium testing of the full-scale pump, CEA planned a joint sodium testing programme for SPX-2 Pump mock-up in cooperation with Jeumont-Schneider. The plan was to test a 1:5 scale mock-up of the SPX-2 impeller in water (at Jeumont-Schneider) and sodium (at Cadarache) to study the impeller performance and confirm the absence of cavitation damage in sodium for the selected cavitation criterion.

As part of the development work towards reducing the lateral dimensions of the primary pump for the EFR reactor, NPSH and cavitation erosion tests were done in sodium and water on the primary pump impeller.

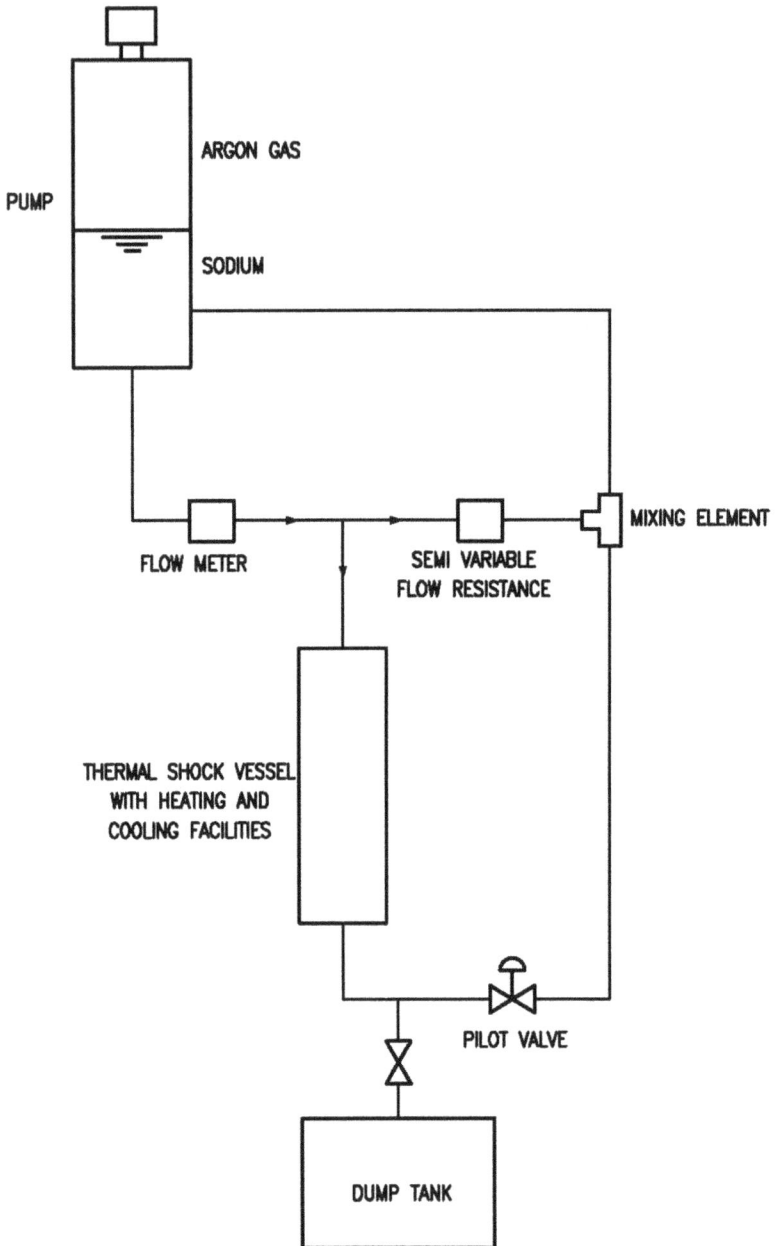

Figure 5.18 Basic schematic of sodium test facility at TNO Laboratory, Apeldoorn, Netherlands [25]. (Fakkel R.H., "Sodium pump development", Sodium-Cooled Fast Reactor Engineering, Proceedings of a Symposium Held in Monaco, 23–27 March 1970, IAEA, Vienna (1970) 343–374. Reproduced with kind permission of the International Atomic Energy Agency (IAEA).)

(vi) Pump test facility in the Netherlands [25]: A small sodium pump test facility was commissioned as early as 1968 to garner experience in the operation of a small sodium pump of capacity 280 m³/h at TNO Laboratory, Apeldoorn (Figure 5.18). The objective of this programme was to gain hands-on experience in the design, fabrication, and testing of sodium centrifugal pumps. During normal operation, the main flow rate in the loop was varied by a semi-variable resistance. The temperature in the circuit was maintained by bypassing a fraction of the flow rate through a cooling vessel to remove the heat generated from the pressure drop in the circuit. Testing under thermal shock (cold/hot) conditions was achieved by injecting liquid from a thermal vessel maintained at a temperature different from that in the loop. The loop was designed for a maximum temperature of 650°C and a maximum thermal shock of 200°C in 3 s. The loop was used to test the 280 m³/h pump in sodium at temperatures from 400°C to 600°C.

5.14 PUMP PERFORMANCE TESTING IN SODIUM – TROUBLESHOOTING EXPERIENCE

It has been established in theory and practice that the in-sodium hydraulic performance of sodium centrifugal pumps can be confirmed by testing in water at room temperature. Water testing has the advantage of low capital and operating cost, as well as ease of operation. However, testing in sodium, although more expensive and challenging, is invaluable for acquiring expertise in the following areas:

(i) The effect of temperature on parts operating with close clearances, such as hydrostatic bearings and wearing rings.

(ii) The effect of thermal shock on pump operation. Thermal shock can result from typical events such as inadequate flow rate compensation in response to power change resulting in reactor shutdown or rare events such as SCRAM. The resulting rapid rate of change of liquid temperature can result in local distortion of parts and affect the smooth operation of the pump.

(iii) The effect of sodium vapour deposition in clearances in the cover gas space of the pump. Sodium aerosol concentration in the argon cover gas space increases exponentially with the temperature of the sodium pool [26]. The condensation of sodium vapour in cooler regions in the cover gas space results in the closure of the clearances, e.g., the clearance between the rotating shaft and the stationary surface in the roof slab of the reactor vessel, and this can interrupt the smooth operation of the pump.

(iv) Convection cells, resulting from density differences, in the cover gas space can result in the formation of convection loops around a portion of the circumference (cellular convection). Such convection loops can produce an asymmetrical temperature differential along the circumference resulting in bowing/tilting of the components (e.g., pump shaft bowing, main vessel tilting) and adversely affect the clearance between the rotating and stationary components (e.g., clearance between the pump shaft and thermal baffles/opening in the roof slab).

(v) Piping reaction on pump nozzles in the case of pumps in the loop type configuration can affect the clearance in the interface between the pump and piping.

(vi) Confirmation of other features such as the efficacy of thermal baffles in limiting the heat transfer to the pump flange, the evaluation of variable speed drive performance and its accuracy in flow control, measurement of bearing loads under different operating states, and conformance of vibration levels with applicable codes.

(vii) Handling of sodium-wetted components and sodium removal. Cleaning sodium from the pump is the first step before the pump is dismantled for maintenance. Sodium removal is a specialised operation, and sodium testing of the pump can provide hands-on experience in cleaning the full-scale component after the tests are completed.

The following paragraphs discuss the experience gained from sodium testing of various reactor pumps.

5.14.1 HNPF [27, 28]

The pump was tested in SS304 loop of 12″ (304.8 mm) size provided with a 12″ (304.8 mm) ball-type throttling valve at the pump discharge and a magnetic flow meter. The system was provided with an expansion tank at the pump suction.

Two major problems during the tests were related to the seizure of the pump shaft and the entrainment of cover gas into the pump. Two seizures of the pump shaft with the aluminium–bronze bushing at the bottom of the radiation shield plug occurred. Investigations revealed that the misalignment of the pump assembly was the possible cause of the seizure. Since the alignment of the pump subassemblies was challenging, the clearance between the shaft and the bushing in the radiation shield plug was increased from 3.1 mm to 6.2 mm to resolve the problem.

Gas entrainment into the pump suction was another problem experienced during testing. Sodium leaking into the free surface pool above the impeller was routed through an overflow line (150 NB) back to the pump suction. It was observed that at flow rates exceeding 5,300 gpm (0.334 m³/s) (as compared to the rated pump flow rate of 7,200 gpm (0.454 m³/s) the pump flow became unsteady due to gas entrainment into pump suction through the overflow line.

Relocating the overflow line to the pump suction was of no avail, and the problem was resolved by ensuring the total submergence of the overflow line inlet. The pump was tested for a cumulative period of 800 h at temperatures of 350°F–1000°F (176.7°C–537.8°C), speeds of 227 rpm–1,135 rpm and flow rates up to 9,000 gpm (0.567 m³/s).

Disassembly and cleaning of the pump were tedious because the shaft was delicate, and there was no provision on the shaft for handling. Also, removing sodium from the wearing ring labyrinth required applying large force, underscoring the need to provide tapped holes for jack screws.

Plugging of instrument thimbles in the inner assembly to prevent the flow of sodium vapour to the driven rotor of the eddy current coupling area was also found necessary. All these modifications were implemented in the reactor pumps.

5.14.2 EBR-II [28–30]

A prototype pump of capacity 5000 gpm (0.315 m³/s) at a head of 107 ft (32.6 m) was tested at temperatures ranging from 700°F (371.1°C) to 900° F at different speeds in a 12" (304.8 mm) dia. sodium test loop. The performance curves plotted at speed of 1750 rpm at 700°F (371.1°C), and 870 rpm at 600°F (315.6°C) showed stable H-Q characteristics with a pump overall efficiency of 76%. During the initial 6,500 h of operation, the pump was operated at 800°F (426.7°C) for four days, at 850°F (454.4°C) for one day, and at 900°F (482.2°C) for one day; no difficulties were experienced during the operation. The pump was also subjected to 100 starts during this period. During the tests, the oxygen concentration in sodium was at saturation level because there was no provision to control oxide impurities in the loop. After 8,500 h of operation, the pump was stopped, disassembled, and inspected. There was no evidence of any corrosion despite operating with saturated oxygen concentration. The hydrostatic bearing surface was free of scoring marks except for an insignificant groove 1/1,000" (~ 25 µm) deep. However, there was shrinkage of bores in the pump casing, which required machining of the hydrostatic bearing journal and wearing ring by 0.006" (152.4 µm) to achieve the desired operating clearances.

The pump was then re-assembled and operated at a very low speed for 1,000 h to check the bearing operation at low discharge pressure. The pump was also subjected to several starts at this low speed. The cycle of tests was then repeated.

The prototype pump was successfully tested for 16,000 h with 250 starts. The tests included 7,000 h at full speed of 1,750 rpm, 3,700 h at half speed, and 5,200 h at 10% speed.

5.14.3 PFR [31–33]

A vertical medium capacity centrifugal sodium pump was manufactured and tested in sodium as part of development work towards designing centrifugal main coolant pumps for the Prototype Fast Reactor (PFR).

The pump was tested in sodium in a 12″ (304.8 mm) loop over a range of speeds between full speed and half speed for about 3000 h, including stopping and starting over 250 times.

The tests indicated: (i) minor cavitation damage in the impeller; (ii) significant sodium vapour deposition in the cover gas space; and (iii) difficulty in dismantling parts immersed in sodium, and provided the necessary information to improve the design.

All three primary sodium pumps were tested in sodium during site commissioning. Primary pump 3 seized when it was re-started after stopping at 320°C for 6–7 h to attend to inspection on ancillary equipment. The pump seized about 25 min. after re-starting. Examination and investigations revealed that the shaft had bowed and seized at the biological shield plug in the cover gas space resulting in contact and additional jamming at the hydrostatic bearing in sodium. Inspection showed that the actual clearance between the plug and the shaft was an order of magnitude less than that recommended because the bore of the shield plug had not been finish machined. Analysis revealed that temperature differential across the shaft resulted in shaft bowing and contact at the plug location and that the stiffness of the hydrostatic bearing was lower than the expected value.

Interestingly, one secondary loop was filled with water, and secondary sodium pump-1 was tested in water. Water testing was conducted to (i) establish pump hydraulic performance reference data; (ii) study the vibration behaviour of the steam generator bundle; (iii) understand the vibration of the secondary piping; and (iv) explore the likelihood of gas entrainment.

The water tests showed that the secondary piping resistance was less than the calculated value, resulting in the pump operating at a flow rate higher than the rated value, thus resulting in cavitation. The secondary circuit resistance was increased, and the pump performance was confirmed. The loop was then drained, dried, and re-filled with sodium, and the pump started at 200°C. The pump was operated satisfactorily at various speeds up to 300°C. However, the pump tripped on thermal overload after about 8–9 hours of operation. Examination showed that the shaft had seized at the hydrostatic bearing location, and analysis indicated the possibility of detachment of the hardfacing of the hydrostatic bearing. The other two secondary pumps were tested in sodium alone.

5.14.4 FFTF [34–36]

During the initial water tests for performance validation, three design changes were made viz., (i) the clearance between the impeller and diffuser was reduced to decrease the flow separation at low flow rates, (ii the size of the supply orifices to the hydrostatic bearing was reduced to increase the bearing stiffness, and (iii) the annular gap between the pump shaft and the sodium displacement chamber was increased and the concentricity improved to reduce eccentric loading on the shaft.

The prototype primary pump was afterwards extensively tested in sodium in the Sodium pump Test Facility (SPTF) at the Energy Technology Engineering Centre (ETEC) in California. The prototype was tested in sodium for 6438 h with 839 h at temperatures above 1000°F (538°C), including 20 thermal transients.

The test programme consisted of four general areas:

(i) Pump assembly and installation – This included fixing thermocouples for temperature monitoring, mounting level probes, and installing/calibrating proximity probes for shaft position monitoring.
(ii) Preoperational check covered operating the bearing and seal lubricating oil system, functional testing of the speed control system, and operating the pump preheating system.
(iii) Steady-state operation.

Transient operation. Preheating experience: The pump was preheated from room temperature to 200 $\pm$ 13.8°C at the rate of 5.6°C/h. The pump shaft breakaway torque increased from the room temperature value of 108–115 N-m to 196N-m at 149°C. Although the temperature was held steady for sufficient time to achieve uniform temperature distribution in the shaft and pump vessel, the breakaway shaft torque increased with a further increase in temperature. As part of troubleshooting, 16 thermocouples were installed in the pump tank at 0.9 m, and 1.2 m elevations above the preheat zone to measure the temperature profile. A cold region was identified, and its temperature was raised using heaters with insulation. However, this resulted in the relocation of the cold zone, and there was no reduction in the shaft breakaway torque. Dial gauges were installed to monitor the radial and axial movement at the pump coupling. The shaft was rotated, torque/dial gauge readings recorded, and the shaft was parked in a different position every hour. When the pump tank temperature was increased, the shaft breakaway torque increased to 305 N-m, and the average temperature differential increased to 54°C. However, when the motor was uncoupled, the shaft torque dropped to 80 Nm, and possible misalignment between the pump shaft and the motor shaft was suspected. Thermocouples were mounted on the hold-down studs of the pump stool to study the likelihood of thermal distortion of the stool. However, no evidence of temperature variation proportional to that required to result in the observed increase in the shaft torque was present.

Extensive measurements were done to determine the shaft's thermally induced permanent and reversible bending, but the permanent shaft runout was within the allowable tolerance for shaft bend.

A study of the pump construction indicated the possibility of convection cells developing in the annular space between the pump tank and the shielding plug that could result in the observed temperature differential. An estimation of the Rayleigh number showed that the actual value was 220, while the critical value was 100, indicating the likelihood of strong convection currents in the annular space. Experimental confirmation was done by evacuating the argon cover gas, maintained at 204.4°C, from the pump tank, which reduced the temperature difference to 28°C. However, when argon was re-introduced into the pump tank, the temperature difference increased to 50°C and 45.9°C, respectively, at the 0.9 m and 1.2 m elevations. Further confirmation of the effect of the heat transfer properties of the cover gas in causing the temperature differential was obtained by replacing argon with helium as the pump tank cover gas. Since helium's properties make it less prone to the formation of convection cells than argon, the temperature difference reduced from 49°C with argon to less than 28°C with helium, and the shaft breakaway torque dropped to 47 Nm. However, helium could not be used in place of argon, so baffles were installed in the annular space to impede the formation of the convection cells.

This exercise resulted in the finalisation of a new parameter to quantify the effect of temperature differential on the bending of the pump shaft, $L\Delta T$, where ΔT = the measured circumferential temperature differential, L = axial length over which the temperature differential exists. A limit of the total $L\Delta T$ measured on the pump was fixed at 3,916 cm-°C.

Vibrations were observed, during steady-state testing, due to vortex shedding at the impeller outlet wearing ring. The problem was resolved by stiffening the wearing ring. During the thermal transient testing to evaluate performance under plant off-normal conditions,, seizure of the hydrostatic bearing occurred due to the densification of the bearing support material. The bearing clearance was increased, and the bearing support heat treated to ensure dimensional stability. The pump was then re-tested and satisfactory performance confirmed under normal and abnormal operating conditions.

5.14.5 BOR-60, BN-350, BN-600 and BN-800 [11, 37]

The prototype pumps of these reactors were tested in sodium in addition to water tests. The objective of sodium testing was to understand the specific issues relating to pumping sodium and the effect of thermal problems on the mechanical behaviour of the pump. The duration of endurance tests in sodium for the various pumps was as follows:

BOR-60: 5,000 hr; BN-350: 3,500 hr; BN-600: 1,500 hr; BN-800 (secondary pump): 1,000 hr.

The tests in the BOR-60 and BN-600 pumps were done using impellers of reduced capacity compared to the actual impeller to simplify the test rigs and reduce the power consumption for the endurance tests.

It should be noted that the cumulative time of testing of the prototype pumps is considerably more (for e.g., BN-600 secondary pump operated in the sodium test facility for more than 20,000 h). However, the subsequent BOR-60 and BN-600 pumps were tested in water only, and their performance was extrapolated to sodium.

In addition to conventional performance testing, the tests also covered operating conditions such as low-speed operation, reversed rotation operation, level variation in pump tank, gas entrainment in pump suction, effect of maximum sodium level and temperature (up to 500°C), characterisation of pump vibration, effect of starts and stops on performance of bearings (the number of pump starts was in the range 100–400) and loss of lubrication and cooling to bearings and seals.

The radial load and axial thrust on the pumps were also measured in the tests. While detailed cavitation tests were done in water only, in the sodium tests, the minimum suction head required to avoid cavitation was maintained, and satisfactory pump operation was confirmed.

The sodium testing exercise also provided valuable experience in sodium removal from the pump. The pumps are the largest complex components to be removed from sodium for maintenance as and when required, and complete draining is essential before transferring it to the cleaning pit. It was observed that draining sodium from the pump with the circuit temperature at 300°C ensures minimum adhering film of sodium on the wetted parts. Another useful measurement that was done after the tests and draining of sodium was the force required to remove the pump with the pump test vessel at 100°C–150°C. For the BN-600 primary pump, the force required to remove the pump was 1000 kN (~100 tons), while the self-weight of the pump was 860 kN (~86 tons); for the BN-600 secondary pump, the corresponding values were 150 kN (~15 tons) and 120 kN (~12 tons) respectively.

5.14.6 MONJU [38]

A full-scale mock-up primary pump was tested in the sodium loop at Oharai Engineering Centre(OEC)/PNC. During the tests, a circumferential temperature difference of 74°C was observed in the cover gas space of the pump tank. It was confirmed through experiments that the natural convection of cover gas in the space between the inner and outer casing of the pump was the reason for the same.

The problem was resolved by installing anticonvection baffles to restrict the convection currents in the region, and this reduced the circumferential temperature difference from 74°C to 10°C.

Table 5.2 gives a concise summary of the various tests carried out on the main coolant pumps of sodium-cooled fast reactors worldwide.

Table 5.2 Summary of tests done on fast reactor primary pumps

Ser	Class of reactor	Reactor	Model	Water	Full-scale Sodium
1	Experimental	EBR-II [28–30, 39]	a	Yes	Yes
2		EFAPP [40]	a	Yes	Yes
3		RAPSODIE [41]	Yes	Yes	Yes
4		BOR-60 [11]	1 : 2.2 scale	Yes	Yes
5		KNK II [42]	a	a	Yes
6		FFTF [33, 35, 43]	a	Yes	Yes
7		FBTR	a	Yes	a
8		JOYO [44, 45]	Yes	Full-scale	Full-scale
9	Demonstration	BN 350 [11]	1 : 4.5 scale	Yes	Yes
10		PFR [33]	1 : 2 scale	Yes	Yes during site commissioning
11		BN-600 [11]	1 : 5 and 1 : 6.6 scale	Yes	Yes
12		Phenix [41, 46]	Yes	Yes	a
13		SNR-300 [18, 19]	a	Yes	Yes
14		MONJU [38]	a	a	Yes
15		SPX-1 [47]	1:4 scale model test in water	Yes	Full-scale testing of rotor assembly with dummy impeller (of same mass and MI) for thermomechanical behaviour of pump rotor under normal and accidental conditions

16		PFBR [8]	PSP: 1:2.75 scale, water Performance test, NPSHR$_{3\%}$, visual test to measure cavity size, paint erosion test, buffed specimen test	Yes. Performance testing, NPSHR$_{3\%}$, testing, inclined rotor performance test	–
17	Commercial	BN-800 [11]	1:4.35 and 1:4 scale	Yes	Yes
18		SPX-2[b] [48]	Cavitation tests on 1/5 scale model in water 1/2 scale model of SPX-2 pump in EPOCA water loop. Pump shaft used as waveguide to detect cavitation	–	–
19		EFR[b] [24]	Primary pump impeller tested in water (in IRIS loop) and later in sodium (in CARUSO loop) to compare NPSH values. Erosion testing of 1360 h duration done in sodium	–	–

[a] data not available
[b] reactor not built

5.15 SUMMARY

Testing of sodium centrifugal pumps is mandatory considering the critical nature of these equipments. These tests are useful in not only establishing the performance parameters of the equipment, but also in understanding the behaviour at high temperature of the pump integrated with the piping system. Although water tests are sufficient to establish the hydraulic performance of the pump, sodium testing is the gold standard in establishing pump performance under the demanding operating conditions envisaged in the reactor.

REFERENCES

1. American National Standard for Rotodynamic Pumps for Hydraulic Performance Acceptance Tests, ANSI/HI 14.6-2016.
2. Stephen Lazarkiewicz and Adam T. Troskolanski, *Impeller Pumps*, Pergamon Press, 1965.
3. Sokolov Kestin and Wakeham, Viscosity of Liquid Water in the Range -8°C to 150°C, *Journal of Physical and Chemical Reference Data*, 7: 3, 1978, pp. 941–948.
4. John C. Crittenden, R. Rhodes Trussell, David W. Hand, Kerry J. Howe and George Tchobanoglous, *MWH's Water treatment: principles and Design*, 3rd edition, John Wiley & Sons.
5. Joanne K. Fink and Leonard Leibowitz, Thermophysical Properties of Sodium, ANL_CEN-RSD-79-1.
6. Yves Lecoffre, *Cavitation Bubble Trackers*, AA Balkema/Rotterdam/Brookfield, 1999.
7. M.C. Zerinvary and E.W. Wagner, Development of an 85,000 gpm (19303 m3/h) LMFBR Primary Pump, Paper no. 115, Proc. of Liquid Metal Engineering and Technology, Vol. 2, Proc. of the Third International Conference held in Oxford on 9–13 April 1984, BNES, London.
8. S.G. Joshi, A.S. Pujari, R.D. Kale and B.K. Sreedhar, Cavitation Studies on a Model of Primary Sodium Pump, *Proceedings of FEDSM02, The 2002 joint US ASME European Fluids Engineering Summer Conference*, July 14–18, 2002, Montreal, Canada.
9. J.F. Gulich, Quantitative Prediction of Cavitation Erosion in Centrifugal Pumps, IAHR Symposium, paper no. 42, 1986.
10. G. Ghia, V. Cela, P. Casalini, G. Mengoli and R. Rappini, Construction, Installation and Sodium Testing of a Test Section Simulating the Rotating Parts of SuperPhénix Primary Pump (PIVOTERIE-1), *Proceedings of Second International Conference on Liquid Metal Technology in Energy Production*, Ed. J.M. Dahlke, pp. 11–10 to 11–15, August 1980.
11. E.A. Belov, F.M. Mitenkov, E.G. Novinski, G.M. Nikolushkin and G.P. Shishkin, Design and Experimental Development of Sodium Pumps, Paper no. 121, BNES, Oxford, UK, 9–13 April 1984.
12. Quarterly Technical Progress Report, NAA-SR-12585, Liquid Metal Engineering Center (LMEC), July–September, November 1967.

13. Quarterly Technical Progress Report, LMEC-68-20, Liquid Metal Engineering Center (LMEC), April–June, October 1968.
14. O.J. Faust (Ed.), *Sodium-Nak Engineering Handbook*, vol. III, Sodium Systems, Safety, Handling and Instrumentation, Gordon and Breach, Science Publishers, Inc., 1978.
15. H. Dieckamp, B. Wolfe and R.W. Dickinson, Development and Testing of Sodium Components, Conf 710901 n.d.
16. J.T. Cochran, G.G. Glenn, W.J. Purcell, R.W. Smith and E.A. Wright, CRBRP Main Sodium Pump Test Experience, Paper no. 113, *Proceedings of Liquid Metal Engineering and Technology*, BNES, London, 1984.
17. C. Dunn and M.J. Gabler, Development of an 85000 gal/min Inducer Pump for LMFBR, Paper no. 113, *Proceedings of Liquid Metal Engineering and Technology*, BNES, London, 1984.
18. K. Mente, H. Reimann, R. Fakkel, M. Heslenfeld, Testing of a Prototype Sodium Pump, INTAT 75.9, February 1974.
19. R.H. Fakkel, C.J. Hoornweg, W.K. Mendte, M. Jansing, H.J. Lameris, J.P. Vroom and M.H. Heslefeld, Development, Design, Construction and Full Scale Sodium Testing of a Prototype Sodium Pump for a LMFBR Power Plant, Paper no. C108/74, Pumps for Nuclear Power Plants, IMechE Convention, University of Bath, 22–24 April, 1974.
20. P.H. Delves and G. Seed, Operating Experiences with the Prototype 0.45m³/s Sodium Pump, Paper no. C104/74, Pumps for Nuclear Power Plants, IMechE Convention, University of Bath, 22–24 April, 1974.
21. Y. Oda, A. Suzuki, S. Kawahara, N. Sagawa, T. Komatsu and Y. Ito, Development of Mechanical Sodium Pump, *Proceedings of Symposium on Sodium Cooled Fast Reactor Engineering*, Monaco, 23–27 March 1970.
22. T. Contardi, L. Rapezzi, P. Le Coz, Y. Tigeot, C. Partiti, M. Zola and P. Denimal, Sodium Test of the Superphénix Full Size Primary Pump Shaft on the CPV-1 Test Rig at ENEA-Brasimone, Paper no. 117, Proc. of Liquid Metal Engineering and Technology, Vol. 2, BNES, London, 1984.
23. P. Courbiere and P. Lecoz, Cavitation in Pumps: Research and Development in France, IWGFR Specialists' Meeting on Cavitation Criteria for Designing Mechanisms Working in Sodium Application to Pumps, Interatom GmbH, Federal Republic of Germany, October 28–29 1985.
24. W. Marth, A Review of the Collaborative Programme on the European Fast Reactor (EFR), *Annual Meeting of the International Working Group on Fast Reactors (IWGFR)*, 1991, IAEA, Vienna.
25. R.H. Fakkel, Sodium Pump Development, *Proceedings of Symposium on Sodium Cooled Fast Reactor Engineering*, Monaco, 23–27 March 1970.
26. A. Sheth, Parametric Study of Sodium Aerosols in the Cover-Gas Space of Sodium Cooled Reactors, Document no. ANL-75-11, Argonne National Laboratory, Argonne, Illinois, March 1975.
27. R.W. Atz, Performance of HNPF Prototype Free Surface Sodium Pump, Report no. NAA-SR-4336, June 1960.
28. P.G. Smith, Experience with High Temperature Centrifugal Pumps in Nuclear Reactors and their Application to Molten Salt Thermal Breeder Reactors, ORNL-TM-1993, September 1967.
29. Orville S. Seim and Robert A. Jaross, Characteristics and Performance of 5000 gpm AC Linear Induction and Mechanical Centrifugal Sodium Pumps,

Proceedings of 2nd United Nations Conference on the Peaceful Uses of Atomic Energy, Vol. 7, pp. 88–93, September 1–13, 1958.

30. Sodium Pump Development and Pump Test Facility Design, Report no. WCAP-2347, Westinghouse Electric Corporation, August 1963.

31. J.M. Laithwaite, L. Bowles and F.M. Delves, Sodium Pumps for Fast Reactors, Paper no. IAEA-SM-130/10, *Proceedings of a Symposium on Progress in Sodium Cooled Fast Reactor Engineering*, Monaco, 23–27 March 1970.

32. K.G. Eickhoff, J. Allen and C. Boorman, Engineering Development for Sodium Systems, paper 5B/5, *Proceedings of the London Conference on Fast Breeder Reactors*, BNES, May 1966.

33. Seed, Bowles, McLeod, Design, Testing and Commissioning of Sodium Pumps for the 600 MW(t) Prototype Fast Reactor, Paper no. C106/74, Pumps for Nuclear Power Plants, University of Bath, 22–24 April, 1974.

34. S.T. Brewer, Three Decades of Breeder Technology in the United States of America, *Proceedings of an International Conference on Advanced Systems and International Co-operation*, Vol. 5, Vienna, 1983.

35. R.W. Atz and M.J. Tessier, High Temperature Testing of a Sodium Pump, Presented at *ASME Winter Annual Meeting*, December 1978, ASME Publication No. 78WA/NE12, 1978.

36. R.J. Dowling, R.L. Ferguson and S.A. Weber, Experience in the Design, Construction and Testing of the FFTF Heat Transport System, *Proceedings of International Symposium on Design, Construction and Operating Experience of Demonstration LMFBR's*, Bologna, Italy, 10–14 April, 1978.

37. Status of Liquid Metal Cooled Fast Reactor Technology, IAEA-TECDOC-1083, April 1999.

38. Toshihiro Odo, Nobutaka Ohkuma, Masaaki Ebashi, Tomoyuki Yamada, Kinya Ogawa and Takuho Miyagi, Design and Construction of Heat Transport Systems of the Prototype Fast Breeder Reactor MONJU, *International Conference on Fast Reactors and Related Fuel Cycles*, Vol. 1, November 1991, Kyoto, Japan.

39. J.R. Davis, G.E. Deegan, J.D. Leman and W.H. Perry, Operating Experience with Sodium Pumps at EBR-II, Report no. ANL/EBR-027, October 1970.

40. T.P. Ross, *Primary Sodium System Pumps in the Enrico Fermi Atomic Power Plant, APDA-309*, Atomic Power Development Associates, Inc., June 1969.

41. W. Raozynski, J.P. Delisle, and J.L. Befree, L'Evolution des Pompes a Sodium de Rapsodie a Phenix et a la Filiere, Proceedings of a symposium on Progress in Sodium Cooled Fast Reactor Engineering held by the International Atomic Energy Agency in Monaco, 23–27 March 1970.

42. K. Mendte, DEBENE Status on Cavitation in Sodium pumps, IWGFR Specialists' Meeting on Cavitation criteria for Designing Mechanisms Working in Sodium Application to Pumps, Interatom GmbH, Federal Republic of Germany, 28–29 October, 1985.

43. R. Buonamici, Fast Flux Test Facility Sodium Pump Operating Experience-Mechanical, Presented at *US/UK Specilaists' Meeting on Sodium Pumps*, San Jose, California, November 1987.

44. Y. Ojiri, A. Suzuki, Y. Ito and S. Sakamoto, Development of Mechanical Sodium Pump, C110/74, Pumps for Nuclear Power Plants, University of Bath, 22–24 April, 1974.

45. M. Kambei and M. Kamei, Cavitation tests for "JOYO" Primary and Secondary main Circulating pumps, IWGFR Specialists' Meeting on Cavitation criteria for Designing Mechanisms Working in Sodium Application to Pumps, Interatom GmbH, Federal Republic of Germany, 28–29 October, 1985.

46. M. Guer, W. Raczynski and G. Keyser, SNECMA-Bergeron Sodium Pumps Development Stage Seen in the Light of the Phenix Experiment, C109/74, Pumps for Nuclear Power Plants, University of Bath, 22–24 April, 1974.

47. P. Courbiere, Superphénix 1 Sodium Pumps: Cavitation and Scale Effects, IWGFR Specialists' *Meeting on Cavitation Criteria for Designing Mechanisms Working in Sodium Application to Pumps,* Interatom GmbH, Federal Republic of Germany, 28–29 October, 1985.

48. W. Marsh, A Review of the Collaborative Programme on the European Fast Reactor (EFR), Report no. IWGFR 83, *Annual meeting of the International Working Group on Fast Reactors,* Tsuruga, April 1991.

Chapter 6

Centrifugal sodium pump instrumentation

6.1 INTRODUCTION

The instrumentation employed to measure performance parameters in centrifugal sodium pumps are distinct from conventional pumps. This is mandated by the high-temperature operation and more importantly from the pyrophoric nature of liquid sodium. The instrumentation employed must therefore ensure that its introduction does not compromise the leak tightness of the system in any way. In test loops and reactor systems molten sodium is used as the circulating medium and being a liquid metal the instrumentation employed often exploits the electrical and magnetic properties of metals.

The parameters that are monitored during pump operation are:

(i) Flow rate.
(ii) Pressure at both suction and delivery.
(iii) Speed.
(iv) Sodium level in pump tank.
(v) Operating temperature.
(vi) Vibrations.

6.2 MEASURING PUMP PARAMETERS

6.2.1 Pump flow rate measurement [1, 2]

There are many established methods for pump flow rate measurement which employ instruments such as orifice meter, venture meter, etc. However, these techniques are not widely used for flow rate measurement in reactor sodium systems/experimental loops because they require penetrations in pipelines, which are potential areas of sodium leak to the atmosphere. Moreover it may often be necessary to provide local heating in the line leading to the transducer to avoid sodium freezing resulting in unhindered leak of sodium to the atmosphere in the event of a crack/failure in the line. Hence, as far as possible, instruments that do not require penetration of the sodium boundary are preferred.

 DOI: 10.1201/9781003460350-6

Magnetic flowmeters depend on the electromotive force (emf) generated during the passage of an electrical conductor (liquid metal) through a magnetic field. These flowmeters are classified as:

a. Permanent magnet flowmeter: This is the most common type of electromagnet flowmeter and is illustrated in Figure 6.1. It is most suited for piped systems because it can be mounted externally on the pipe, and no penetration of the sodium boundary is required. The emf, E, produced in the liquid metal (sodium) is given by E = BVd volts where B, in Tesla, is the flux density in the air gap of the magnet, V, in m/s, is the velocity of sodium and d is the duct inside diameter, in m.
   The emf, E in terms of the flow rate, Q, in m^3/s, is given by:

$$E = 4BQ/(\pi d)\,V.$$

   The terminal voltage is less than the emf produced because of shunting due to the conductivity of the duct wall and shunting due to conductance through the liquid metal at the end regions where the magnetic field is low. Wall shunting is reduced by using minimum wall thickness and maximum material resistivity compatible with liquid-metal pumping. Similarly, non-magnetic duct material (stainless steel for sodium application) is used to avoid the shunting of flux around the liquid metal. Wetting of stainless steel by sodium is poor during the initial period of operation, especially at temperatures below 300°C, and this can result in erratic output due to fluctuating electrical contact.

b. Electromagnetic flowmeter: In an electromagnetic flowmeter the magnetic field required is produced using an electromagnet. As the magnetic flux density of an electromagnet is proportional to the current passing through the coils, the power supply used should be of

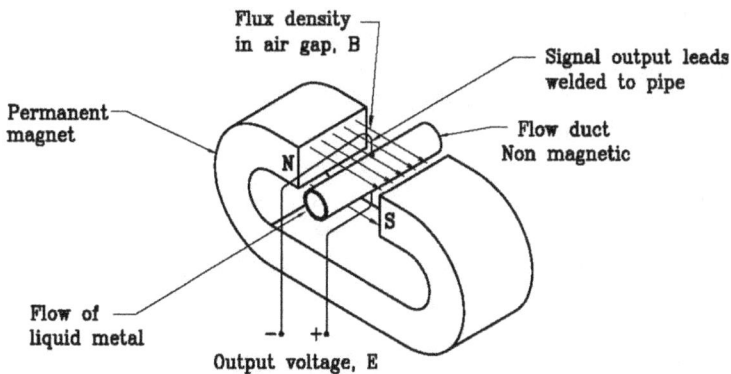

Figure 6.1 Permanent magnet flow meter.

sufficient capacity to ensure a current of constant magnitude. This is critical to guarantee the desired accuracy in flow rate measurement. The electromagnetic flowmeter has two advantages: (i) the magnetic flux density can be varied by controlling the current supplied and in some cases this is favourable in meeting the configuration and weight requirements; and (ii) in applications using large flowmeters where the magnetic field extends over several diameters to reduce non-linearity due to flux shift, an electromagnetic flowmeter is preferred to reduce magnet size and weight. However, the disadvantages of this type of flowmeter are: (i) it requires a well-controlled and continuous source of power supply; (ii) it weighs more than the permanent magnet flow-meter if the weight of the power supply is taken into account; and (iii) it is affected by high coil temperature and the effect of radiation on coil insulation.

c. Eddy current flowmeter: The eddy current flowmeter belongs to the category of alternating current electromagnet flow meter. This type of flowmeter exploits the change in the eddy currents induced in sodium due to variations in flow velocity. In its simplest form, the sensor consists of a centrally located primary winding flanked on either side by identical secondary windings that are differentially connected. The primary and secondary coils are wound on an iron former (bobbin), and the assembly is inserted inside a thimble made from a non-magnetic material. The thimble protects the sensor from the hot sodium that flows around it. The primary winding is energised using a constant current, AC source, and the secondary windings are balanced such that the flow meter output is zero when the probe is in static sodium. In the presence of flowing sodium, the change in the eddy currents induced in sodium produces a distortion of the magnetic field resulting in a difference in the electromotive force (emf) in the two secondary windings. This difference in emf is proportional to the velocity of flow.

The probe output with differentially connected secondary windings is sensitive to sodium temperature because the resistivity of sodium and consequently the eddy currents induced in it change with varying temperature. A temperature-independent probe output is obtained by measuring the secondary outputs independently and computing the ratio of the difference in the outputs to the sum of the outputs.

Figure 6.2 illustrates the eddy current flow meter.

d. Ultrasonic flowmeter: The ultrasonic flowmeter computes the flow velocity using the measured time for a sonic pulse to travel a known distance in the flow field. This flowmeter belongs to the category of Transit-time flowmeters. In this flowmeter, an ultrasonic transmitter is mounted midway between two receivers. Both the transmitter and the receivers are mounted flush with the pipe wall. The instrument is used either in the pulse or continuous modes. In the pulse mode, an ultrasonic pulse is emitted by the transmitter, and the difference in

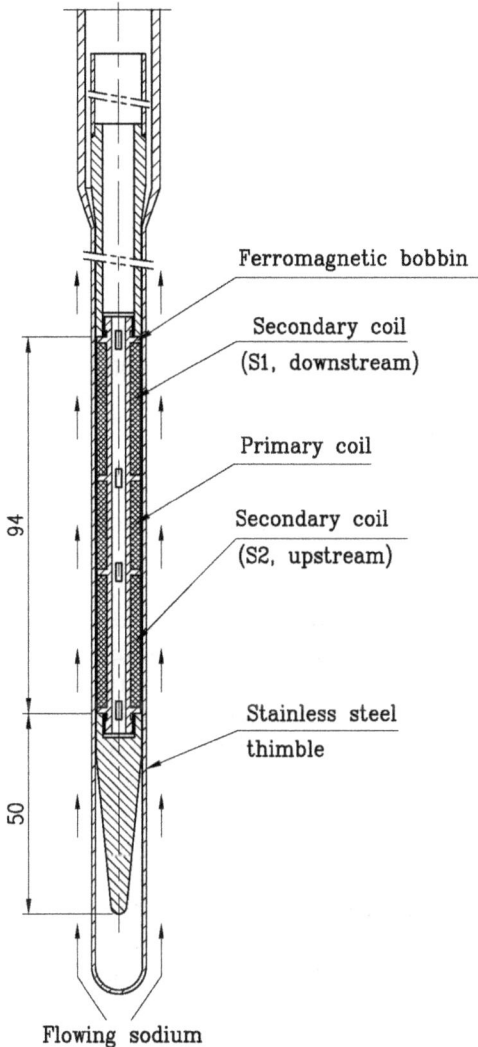

Figure 6.2 Eddy current flow meter.

the time taken by the pulse to reach the upstream and downstream receivers is recorded; in the continuous mode, the transmitter continually emits energy into the flowfield, and the difference between the phase angles of the receiver signals is a measure of the transit time. The transit time thus measured is used to compute the flow velocity using the formula [2].

$$V = \frac{C^2 \Delta t}{2d \cot \theta}$$

where:
   V is the flow velocity
   C is the velocity of sound in the liquid
   Δt is the time difference between the signals at the receivers
   D is the inside diameter of the pipe
   θ is the angle between the flow direction and the line connecting the transmitter and the receiver

6.2.2 Pump pressure measurement [1, 3, 4]

In conjunction with flow rate measurement, pump pressure measurement is essential to obtain the operating point of the pump at the speed of operation.

Methods employed to measure pump pressure without penetration of the pipeline (sodium boundary) are:

(a) Pressure pot: Most pressure-measuring instruments in sodium use an elastic element such as a diaphragm, bellows, or Bourdon tube to measure pressure. However, the pressure pot uses the displacement of a column of sodium against gravity. In this method, a separate pipe (pressure pot) is welded to the suction and discharge lines of the pump and initially isolated from the system using valves. The pots are also communicated with the cover gas system through a separate header. Each pot is provided with mutual inductance type, discrete level probes to detect operation low level, operation high level, and vessel high level, and a pressure gauge to measure the cover gas pressure. After the pumps are started, the valves connecting the pump suction and discharge lines to the respective pressure pots are opened, and the cover gas pressure is adjusted to stabilise the liquid column in each pressure pot at the operating low level. The gauge pressure at the bottom of the pot in the suction/discharge line is then the sum of the cover gas pressure and that resulting from the liquid column in the respective line. Thermocouples are provided in the pressure pot to monitor the temperature of sodium in the pot.

However, this method is cumbersome and used only in experimental loops. Figure 6.3a illustrates the Pressure Pot and Figure 6.3b is a line sketch of the pots mounted on the suction and discharge lines.

The commonly used pressure-measuring instruments employ a pressure gauge with an elastic sensing (deflecting) element, such as a diaphragm, bellows, or Bourdon tube. Here, the sensor is located outside the sodium pipeline and connected to the system through a short length of tube (standoff). This arrangement can result in errors in measurement due to: (i) the effect of the tube on the response time, which is influenced by the geometry and is negligible if the length is kept small; and (ii) the height of the liquid column. This is accounted for by considering

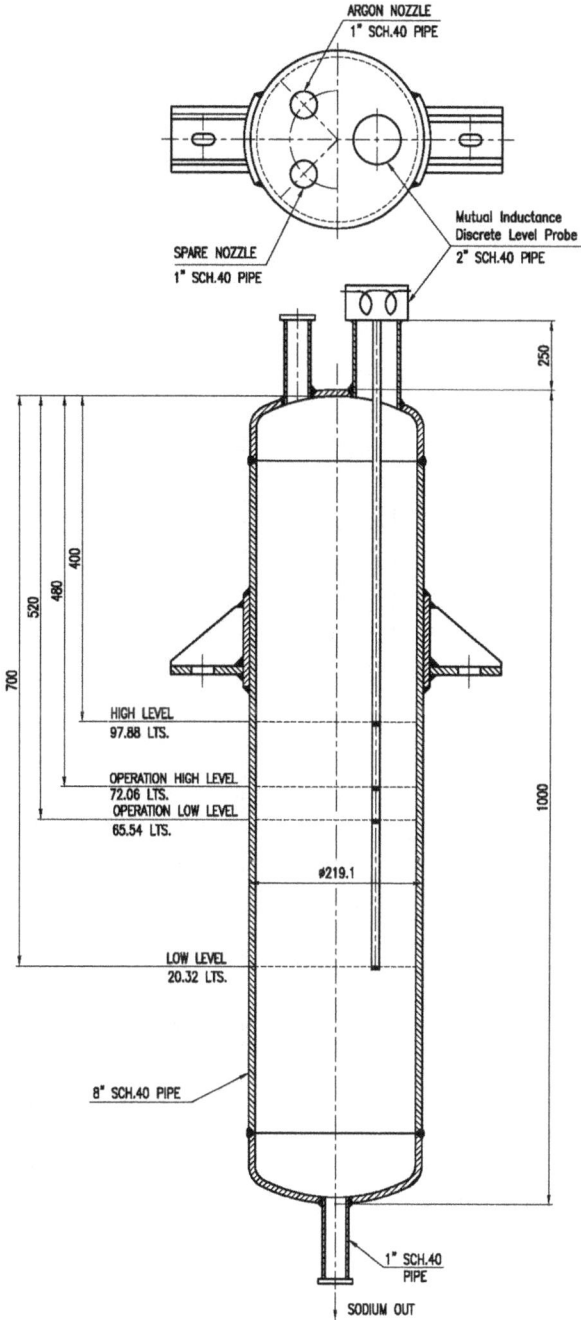

Figure 6.3 (a) Pressure pot.

(Continued)

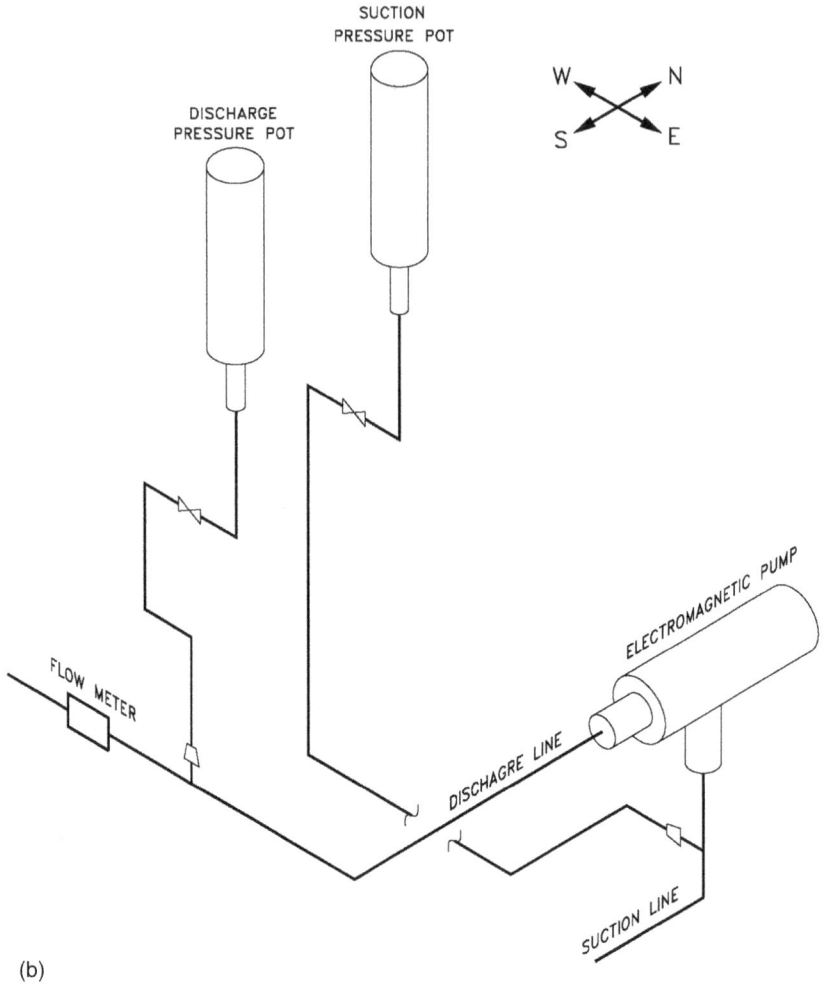

Figure 6.3 (Continued) (b) Pressure pots on suction and discharge lines of the pump.

the elevation difference between the sensing element and the measurement location.

(b) Unrestrained diaphragm pressure sensor: Figure 6.4 shows a pressure sensor that has an unrestrained diaphragm integral with the housing. The deflection of the diaphragm under sodium pressure is sensed by a variable impedance transducer located in close proximity. An oscillator and demodulator unit converts the sensor output to a DC output signal.

(c) Na–K-filled diaphragm pressure sensor: The sensor here consists of a Na–K-filled tube with a multi-ply Inconel diaphragm located within an Inconel housing. The Na–K-filled capillary tube terminates in a

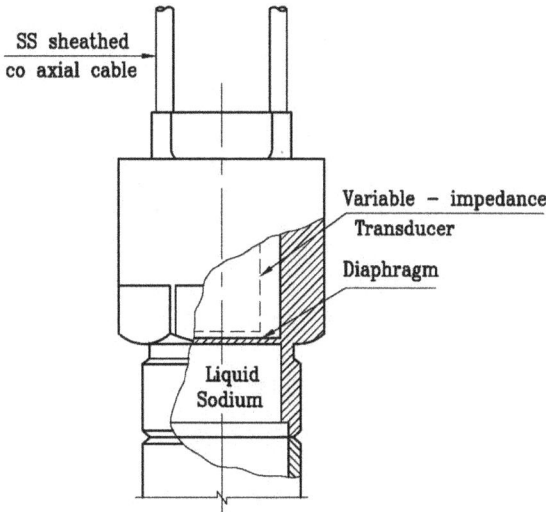

Figure 6.4 Unrestrained diaphragm sensor.

Bourdon spring. The transmitter is a pneumatic force balance type. In operation, the pressure exerted by sodium on the diaphragm is transmitted hydraulically from the hot sodium region (by the Na–K in the capillary tube) to a Bourdon spring that terminates in the transmitter case at room temperature. Na–K, a liquid at room temperature, prevents freezing in the capillary tube. The movement of the Bourdon spring acts on a force beam that positions a relay valve pilot to produce a pressure change. The pressure change causes a bellows to exert an opposing force on the force beam. The force required to balance the Bourdon spring force measures the sodium system pressure acting on the diaphragm sensor. Figure 6.5 is a schematic of the Na–K-filled diaphragm pressure sensor.

(d) Bellows pressure gauge: A bellows allows larger displacement than a diaphragm of the same diameter, and this characteristic is exploited in transducers that convert the resulting displacement into an electrical output signal. Figure 6.6 illustrates the bellows pressure gauge. Liquid sodium pressure change causes a variation in the compression of the bellows resulting in a change in the restriction to gas flow rate at the nozzle seat. This variation in gas flow rate through the orifice produces a change in the back pressure, which is converted into electrical output by a pneumatic relay.

(e) Spring-balanced bellows pressure sensor with differential transformer: Figure 6.7 illustrates this type of pressure sensor. In this device, the sensor is Inconel bellows enclosed in Inconel housing, while the transmitter is a differential transformer. The deflection of the bellows under sodium pressure is transmitted through a spring-restrained push rod to

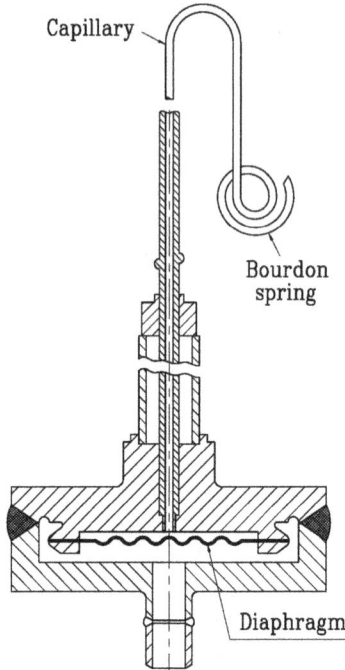

Figure 6.5 Na–K filled diaphragm pressure sensor [3]. (Re-printed with permission from W.R Miller, High Temperature Pressure Transmitter Evaluation, Report no. ORNL-2483, Oak Ridge National laboratory, May 1958).

displace the core of the differential transformer. The spring limits the movement of the push rod and permits span adjustment. The spring stiffness is ~50 times that of the bellows to minimise the effect of variation in the bellows' stiffness with temperature. The fins upstream of the spring ensure that the temperature of the spring is close to room temperature. The output of the differential transformer is calibrated in terms of the sodium pressure.

6.2.3 Pump tank level measurement

Level probes are employed to measure the sodium level in the pump tank. Invariably, the high electrical conductivity of sodium is exploited for level measurement. Level probes are classified as contact or non-contact probes and discrete or continuous probes.

As the name suggests, contact-type probes are in direct contact with liquid sodium. Therefore, the issues of sodium compatibility, high-temperature strength, and wetting of the probe by sodium are critical for the functioning

Input line

Output line

Inconel nozzle
welded to

Type 316 SS screw

Inconel bellows and

Nozzle seat

Process fluid

7"

2 3/4"

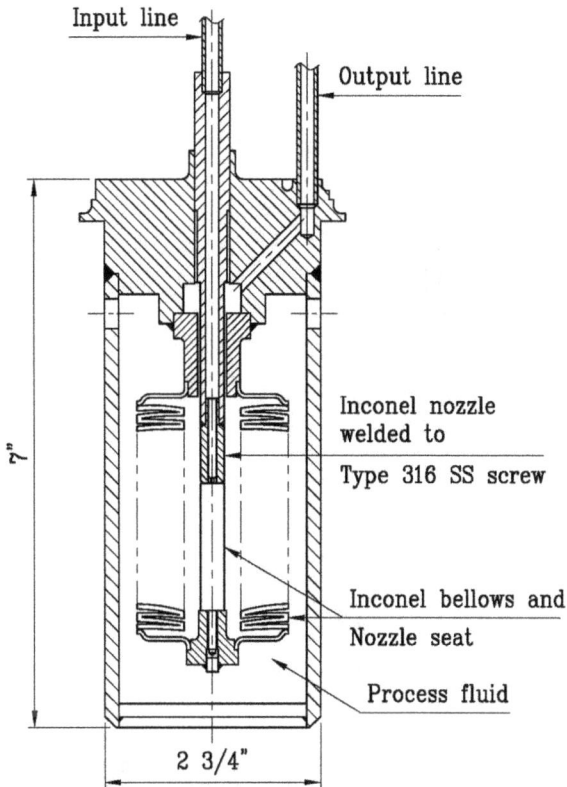

Figure 6.6 Bellows pressure gauge [4]. (Reproduced with kind permission of Argonne National Laboratory, managed and operated by UChicago, Argonne, LLC, for the U.S. Department of Energy under Contract No. DE-AC02-06CH11357).

of these probes. Non-contact probes are not in contact with sodium because they are mounted inside a pocket/thimble or outside the sodium vessel.

Discrete or discontinuous level probes are used to measure the level at a particular elevation in contrast to continuous level probes that can provide level measurement over a continuous range of values. However, discrete probes offer better resolution compared to continuous level probes; multiple discrete probes spaced closely can also mimic a continuous level probe.

6.2.3.1 Contact-type level probe

(a) Spark plug type: The spark plug-type level probe is the simplest type of level probe. It is a direct contact, discrete-type probe, and consists of a conducting rod encased in an insulating sleeve and projecting into the vessel space. The rod comprises one electrode of an electric circuit,

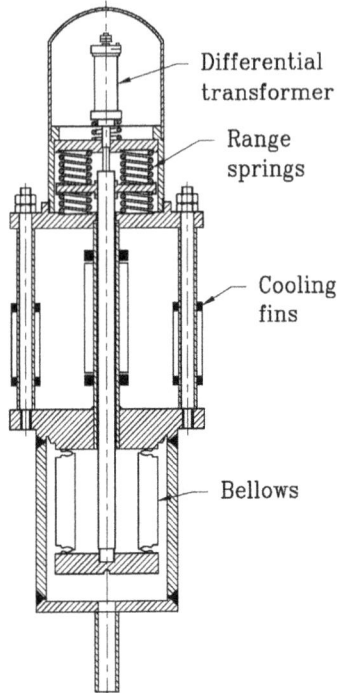

Figure 6.7 Spring-balanced pressure sensor with differential transformer [3]. (Re-printed with permission from W.R Miller, High Temperature Pressure Transmitter Evaluation, Report no. ORNL-2483, Oak Ridge National laboratory, May 1958).

of which liquid sodium is the other. When the sodium level rises and contacts the rod, the electrical circuit is closed, and the indicator is activated. Although the design is simple, the probe suffers from the disadvantage that spurious signals can result from shorting of the circuit due to sodium vapour deposition on the insulator.

(b) Resistance-type probe: This probe consists of a copper rod within a stainless-steel tube (Figure 6.8). The copper rod is insulated along its length using ceramic beads, and the lower end of the copper rod is brazed to the bottom of the tube. The probe is energised using a constant current AC source. The resistance of the probe in the dry condition is that of the copper rod and the SS tube in series. The circuit is shunted when sodium contacts the probe resulting in a step reduction in the voltage across the probe, indicative of the sodium level.

The proper functioning of resistance-type level probes depends on the shorting of the tube along the length covered by liquid sodium. The probe is effective when the contact resistance between sodium and the tube is low and constant, and there is good wetting of the tube

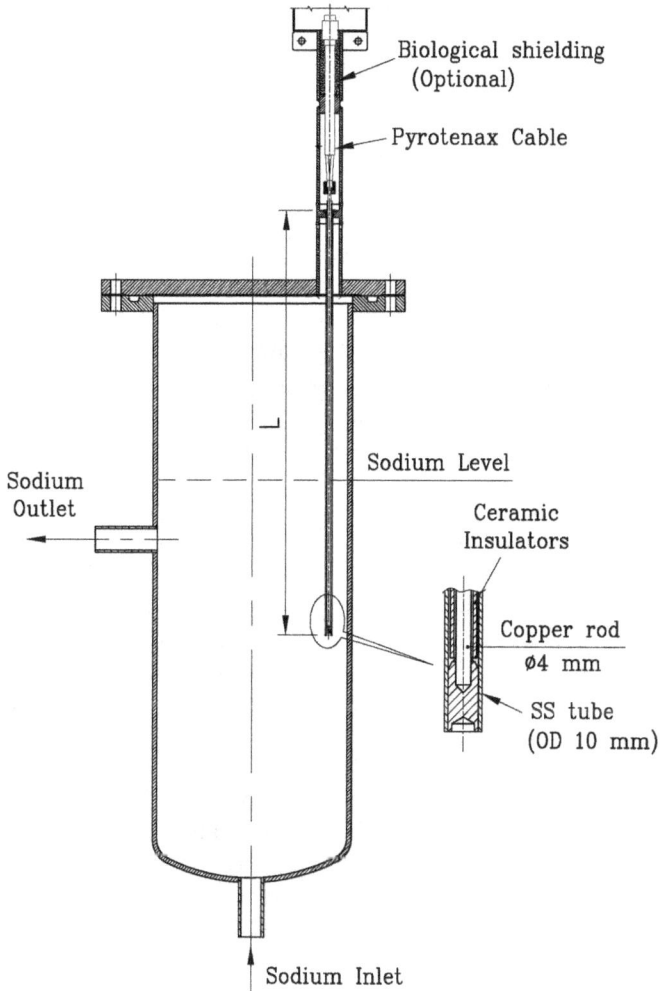

Figure 6.8 Resistance-type probe.

by sodium. Wetting by sodium depends upon factors such as temperature and time of operation, tube surface roughness, and the nature of the tube material. Good wetting is normally achieved above 300°C after about an hour of operation. Wetting is improved by avoiding polished surfaces and using rough surfaces (e.g., with a sandblasted surface). Another factor that affects performance is the resistance per unit length and thickness of the tube. The tube material is to have high resistance per unit length, and the thickness of the tube is to be as small as is feasible from considerations of strength. This probe falls under the category of direct contact, discrete type.

6.2.3.2 Non-contact-type level probe

When an alternating current energises a coil, the changing magnetic field induces a current in the closed conducting path surrounding the coil. The magnetic field due to these eddy currents opposes the changing magnetic field, thus reducing the effective inductance of the coil. The inductance gauge uses this principle for level measurement. In its simplest form, the probe consists of a coil, energised by a constant current AC source, inside a stainless-steel thimble or pocket immersed in the liquid where the level is to be measured. Variation in the liquid level around the probe causes a corresponding change in the coil's inductance, which indicates the level change. The probe response is a function of parameters such as the resistivity to wall thickness ratio of the thimble, the frequency of the AC, the ratio of coil diameter to thimble outside diameter, and the coil length to diameter ratio. There are two types of inductance probes: self-inductance and mutual inductance.

(a) Self-Inductance probe: Figure 6.9 shows a schematic of a self-inductance probe that gives a continuous measurement of sodium level. A single coil of sufficient length to cover the measurement range is connected to an AC source, and its impedance is measured. A reference coil provided for temperature compensation is located out of the range of the liquid sodium but in the proximity of the primary coil so that it experiences nearly the same temperature as the primary coil. The reference coil is connected to the primary coil in a bridge circuit and compensates for the effect of temperature on the probe output. The resolution of the probe is proportional to the full-scale reading or coil length, and therefore the uncertainty in the output is significant for large-range probes. This disadvantage is overcome using a step level gauge [5]. Here, the coil length is finalised based on the required resolution, and an even number of coils are stacked inside the thimble/pocket to cover the range of interest. The coils are connected to a bridge circuit using a selector switch, and the thimble is inserted into the sodium pool. The coils are connected in sequence until the liquid/gas interface is identified. The stack position of this coil gives the reference level, and this value added to the indicator reading gives the sodium level in the pool.

(b) Mutual Inductance (MI) probe: The MI probe consists of a primary and a secondary coil wound in a bifilar fashion on a stainless-steel former encased inside a stainless-steel thimble/pocket.

Mineral insulated cables with a copper conductor that can withstand high temperature and radiation environments are used for the windings. The primary coil is energised with a constant AC source resulting in induced emf in the secondary coil. The presence of sodium outside the tube reduces the secondary emf due to the eddy currents

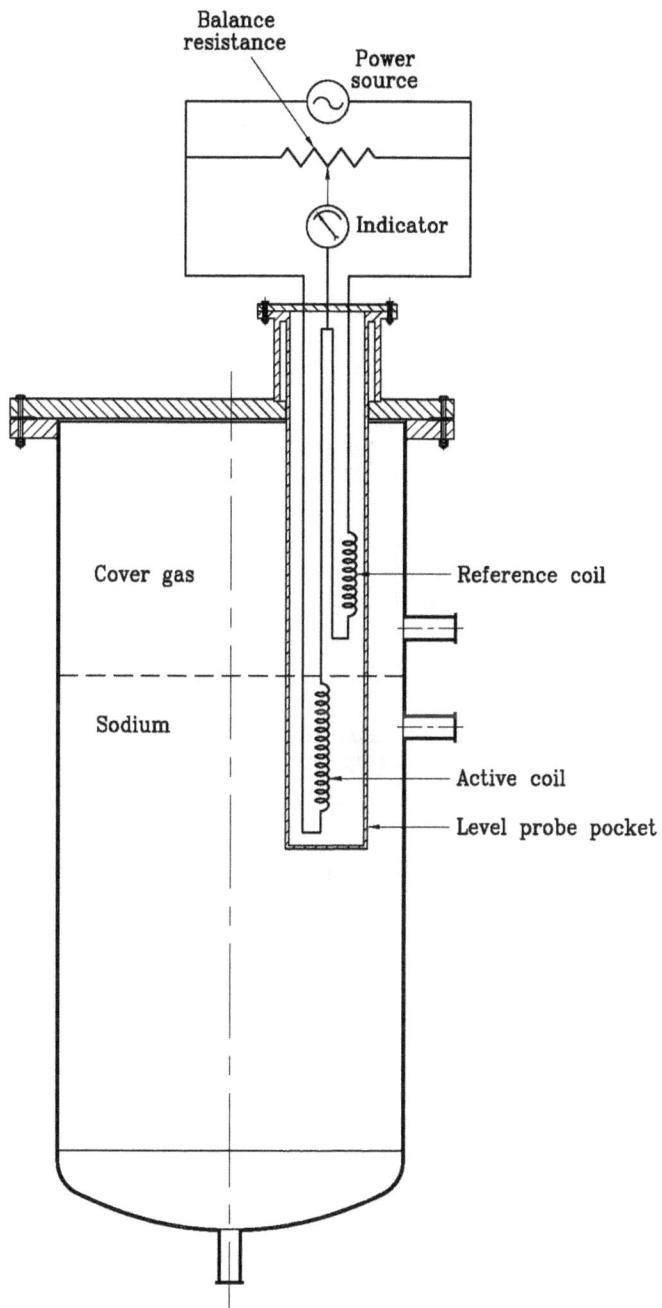

Figure 6.9 Self-inductance probe.

in sodium and the reduction in the secondary voltage is proportional to the sodium level outside the thimble. The effect of temperature on the sensor output is compensated by the electronics. Figure 6.10 gives a schematic of the continuous MI probe. A discrete sensor working on the same principle uses primary and secondary coils of much smaller lengths (~25 mm). Therefore, the temperature compensation in the discrete sensor is negligible, making the electronics more simple.

(c) Differential-Transformer Level Gauge: The differential-transformer level gauge (aka dipstick level gauge) exploits the voltage imbalance, in the secondary coils, resulting from the movement of the trans-former through the cover gas-liquid sodium interface (the imbal-ance results from the significant difference in electrical conductivity in the two mediums). This gauge is, therefore, a discrete-type sensor. Figure 6.11 is a schematic of the probe. The head of the sensor con-tains one primary and two secondary coils enclosed inside a stain-less-steel pocket/thimble. The primary coil is energised by a constant AC source. The secondary coils are wound in series opposition (dif-ferentially connected), and balanced so that the resulting voltage is zero without sodium outside the thimble/pocket. During the lowering

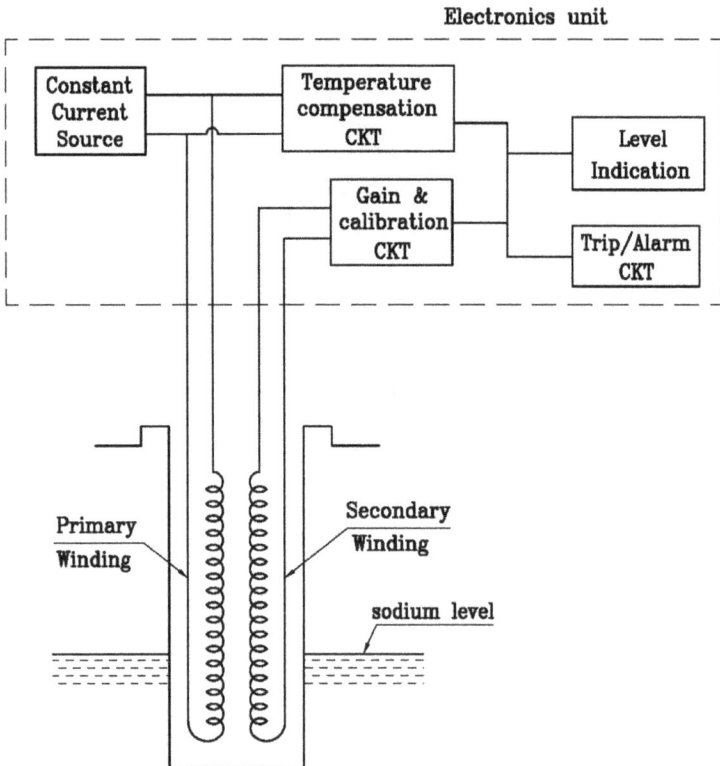

Figure 6.10 Schematic of a mutual inductance probe.

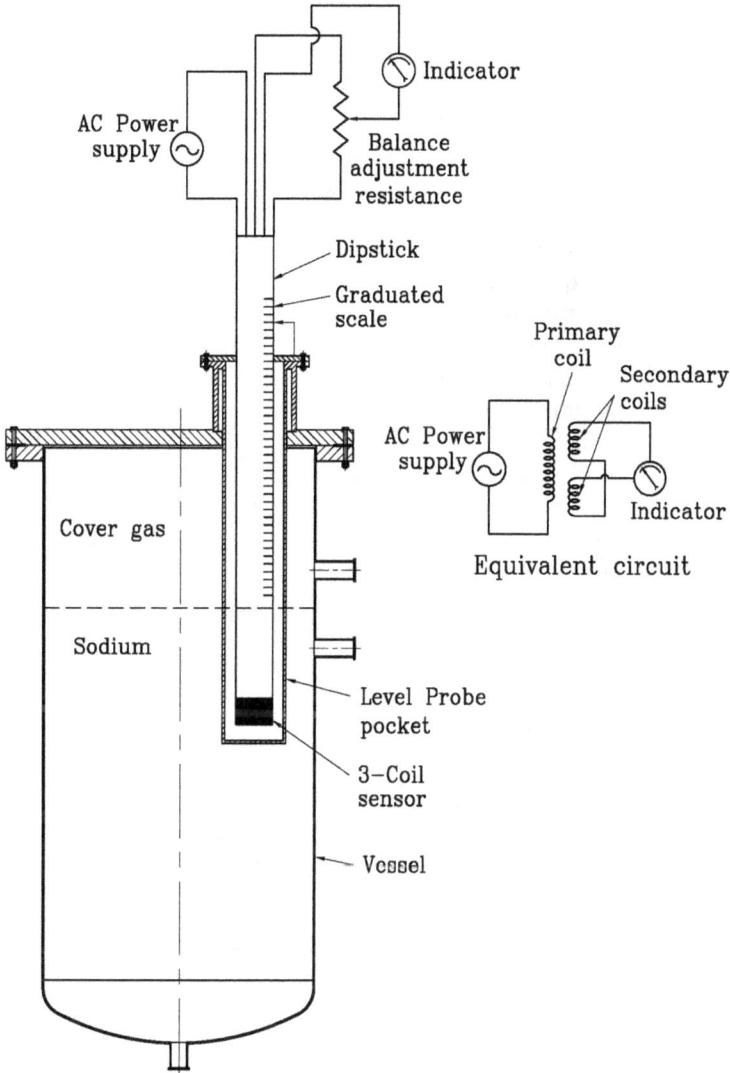

Figure 6.11 Differential-transformer level gauge (dipstick level gauge).

of the probe into the sodium pool, an imbalance between the second-ary voltages occurs when the lower secondary coil comes into contact with sodium. The imbalance is maximum when the liquid/gas inter-face is midway between the two secondary coils and becomes zero (i.e., balance is restored) when both the secondary coils are immersed in sodium. The sodium level in the vessel is obtained using a gradu-ated scale alongside the probe as a reference. The effect of sodium temperature on the probe reading is negligible. The upper limit of the frequency of the primary coil AC is determined by the resistivity and wall thickness of the thimble/pocket and the lower limit from sensitivity considerations. In cases where the probe is to be used at high operating temperatures, ceramic-insulated nickel-coated copper wires are used as the winding material. In such cases, the measurement can be done rapidly using the probe manually (like a dipstick).

(d) RADAR level probe: The RADAR level probe uses the time taken for the reflection of microwaves from the sodium-free surface to estimate the level of liquid sodium in the tank; the product of the wave speed and half the transit time being a measure of the liquid level from a fixed reference point. Figure 6.12a is a sketch of an extended horn antenna probe, and Figure 6.12b is a schematic of the arrangement for level measurement. The RADAR probe has the following advantages:

(i) Overhead space is considerably lower than that for resistance or mutual inductance probes, and handling is easier.

(ii) No temperature compensation is required.

(iii) The probe does not contact sodium; no special cleaning is necessary after removal.

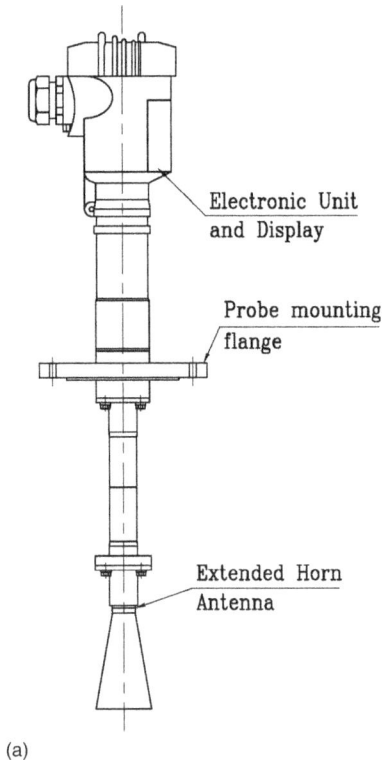

Electronic Unit and Display

Probe mounting flange

Extended Horn Antenna

(a)

Figure 6.12 (a) RADAR Level Probe.   (Continued)

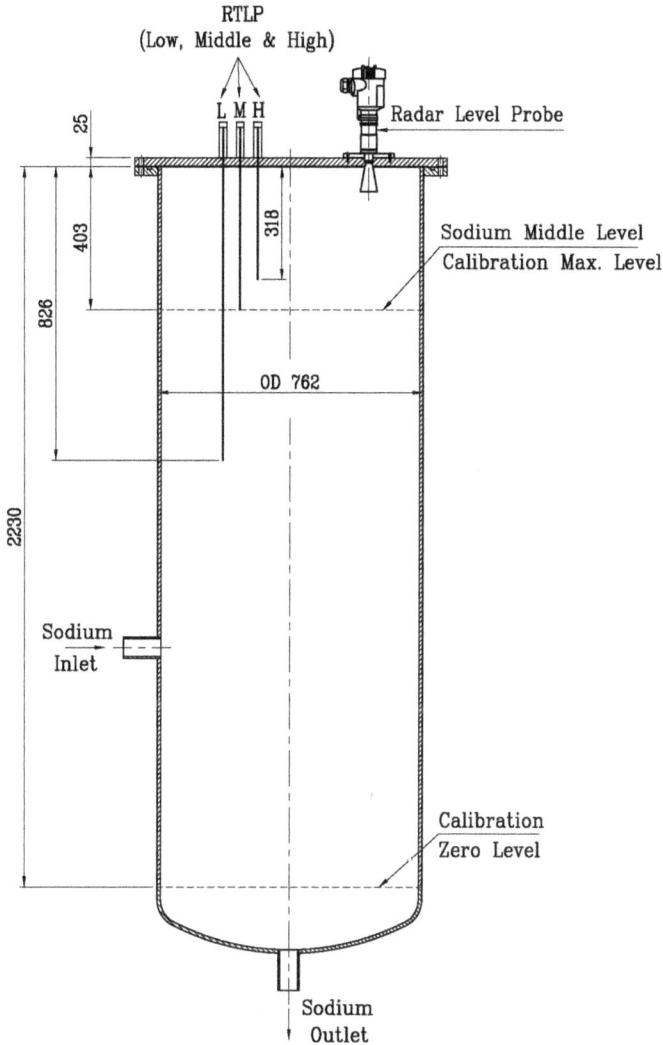

Figure 6.12 (Continued) (b) Level monitoring use RADAR level probe.

6.2.4 Pump speed

Pump speed is measured using a tachometer using the location of the driven shaft near the coupling with the motor.

6.2.5 Sodium temperature

The temperature of sodium in the pump tank is measured using thermo-couples (K type) spot welded to the tank outside surface or using thermowell mounted on the vessel top flange. The thermowell consists of a thermocouple enclosed in a pocket/thimble. The pocket is surrounded by sodium, and it protects the thermocouple inside from coming into direct contact with sodium.

6.2.6 Pump vibration

The vertical rotor assembly of sodium pump is supported at the bottom (inside sodium) by a hydrostatic radial bearing and at the top (outside argon cover gas) by oil lubricated radial sleeve bearing and axial thrust bearing. Vibration measurement is done using accelerometers mounted at the top bearing housing in axial and radial directions and is useful in early diagnosis of pump performance degradation.

6.3 FLOW RATE/CAPACITY REGULATION IN SODIUM PUMPS

In conventional pumps, capacity regulation is achieved using valves. Capacity regulation using valves is avoided in sodium systems because valves are potential leak sources. Instead, flow rate is controlled by varying the speed of the pump using a variable speed drive. In addition to eliminating the possibility of a sodium leak (from valve failure), this method has the added advantage of avoiding energy loss in the valve during throttling. Sodium pumps are designed to permit speed variation from 20% to 100% of the rated speed.

6.4 ONLINE CONDITION MONITORING OF PUMPS [6]

An array of technologies is used for the online performance monitoring of reactor pumps. Although the technologies given below are those that have been employed in light water reactors, they are equally valid for pumps of fast reactors. These include vibration spectral analysis, acoustic emission (AE) analysis, and motor current, or power analysis. Pump vibration can result from a host of hydraulic or mechanical sources and signature analysis of vibration data is invaluable in identifying the source of vibrations. The top hydraulic phenomena responsible for excessive pump vibration include cavitation, flow recirculation, hydraulic radial/axial thrust, gas entrainment, and pressure pulsations. The primary mechanical sources of vibration include degradation of balancing, misalignment of pump and driver (e.g., due to thermal expansion, pipe strain, etc.), improper selection of seals, and bearing failure.

Expert systems are useful in online performance monitoring towards predictive maintenance because it assists in the reduction of occupational radiation exposure during continuous performance monitoring of pumps and other rotating equipment, thus facilitating maintenance as and when required instead of routine procedural maintenance, which often may not be necessary. In the mid-1980s, projects initiated by the Department of Energy (DOE), USA and Electric Power Research Institute (EPRI), USA to demonstrate the use of expert systems for the online diagnosis of problems in rotating machinery in nuclear power plants established the basics of present-day expert systems.

The advantage of an expert system is that it is logical and objective and will not overlook an abnormal condition. It is also beneficial, especially when more than a single fault is responsible for the anomalous condition and several fault patterns are present in the spectral signature. A weakness of the system is that the programmed decision tree is wholly dependent on the software designer and on the accuracy of the database used.

Expert systems that have been used in light water nuclear plants include:

(i) DIAPO (*Diagnostic des Pompes*): This is an expert system used by *Electrcite de France* (EdF) for fault diagnosis at the incipient stage and root cause analysis of failures in coolant pumps of pressurised water reactors (PWR). Parameters such as bearing temperature, relative shaft position, pump seal flow rate, and vibration data such as synchronous, spectral, shock, and resonant frequency data are monitored and analysed. In addition, primary system parameters such as flow rate, pressure, temperature, and power are also monitored. A total of about 200 parameters are monitored and analysed to diagnose the health of the main coolant pump. This program was implemented by EdF in collaboration with the pump manufacturer Jeumont industries.

(ii) PSAD (*Poste de Surveillance et d'Aide au Diagnostic*): EdF used this integrated system for online monitoring and diagnosis of problems in major components/systems of nuclear power plants, such as main coolant pumps, primary circuits, and turbine generators. Each monitored component is connected to a surveillance system through its instrumentation, and in the case of events, the local station displays an online alarm and generates a preliminary diagnosis. Both local and national stations can then access the plant's operation database through a computer network and carry out high-level diagnostic tasks.

(iii) COMOS (Condition Monitoring system) was used in the Grafenrheinfeld KWU pressurised water reactor to monitor the pump shaft vibration and provided valuable data when the shaft of the primary coolant pump broke in December 1986; shaft vibration was monitored using two eddy current sensors mounted in x and y directions. It was observed that shaft vibration amplitude increased over 150% during the last 48 h prior to shaft rupture. The sensors in the system included inductive absolute displacement sensors on the top of the reactor vessel, relative displacement sensors on the pump housing and the hot leg near the steam generator, neutron noise sensors for safety instrumentation, and piezoelectric sensors in the inlet and outlet pipes. The array of sensors was used for periodic measurements, and their output was monitored for deviations from reference signatures. The vibration analysis was based on power spectral densities and correlation functions of selected pump signal pairs. A systematic measurements exercise was done during the pre-operation and plant operational phase to study the spectra in the frequency range of interest. The shaft rupture experience at the Grafenrheinfeld plant increased the confidence in identifying the presence of a crack in a vertical centrifugal pump shaft. The major takeaways regarding shaft crack prediction were that the trend analysis of the amplitude of rotation-specific frequency components is relevant, crack propagation can develop in a period ranging from several hours to several months, and pump operating parameters (e.g., head, temperature, sealing water temperature and pressure) may strongly affect the vibration signals.

(iv) Loose Part Monitoring Systems (LPMS) have been used in German nuclear power plants for condition monitoring of components and online degradation diagnosis. The system analyses acoustic signals with detailed burst data interpretation and trending. Acoustic signals from 12 German nuclear plants provided the database for identifying acoustic signal signatures related to components of the primary system. This system helped detect:

(a) The detachment of a mounting screw and loosening of the impeller cap in the main coolant pump of a PWR.

(b) The weld failure of the cover ring of a pump in a BWR. Unusual acoustic noise at high speeds and burst signals were recorded during reactor startup. Analysis of the data helped find the cause and prevent significant damage.

(v) In the Barsebäck nuclear plant (BWR) in Sweden, a vibration monitoring programme was used to monitor the performance of a wide range of pumps. As many as 12 measurement points per pump and motor set were selected with sensors in horizontal, vertical, and axial directions. In some cases, sensors were mounted on the pump body and the inlet pipe to detect cavitation. Apart from vibration amplitude measurement, spectral analysis was done and classified into seven frequency bands. Process parameters such as flow rate, pressure, and temperature were also analysed. Extensive cavitation damage of the inducer and impeller of the main condenser pump was detected using this programme.

(vi) TEPCO (Tokyo Electric Power Company) sponsored a research project in Japan on condition monitoring methods and systems following a major problem in a circulating water pump in one of its BWR plants. The project resulted in installing eddy current sensors to measure shaft deflection on all the circulating pumps and accelerometers on the pump casings to detect impeller movement, if any, due to cavitation. In 1993, TEPCO decided to monitor the health of safety-related pumps by collecting, analysing, and trending vibration spectral, acoustic, and process information data on a monthly basis. The types of analysis include frequency analysis, wavelet analysis, cascade spectral analysis (e.g., waterfall plots), single or multiple parameter historical trending, and comparison.

(vii) MAINS (Maintenance Support Expert System for Rotating Machines is an offline expert system developed by Toshiba. Although not documented, it probably interfaces with the software developed for TEP-CO's Toshiba pump monitoring system. This software uses data provided by plant personnel to generate a cause–effect matrix that is used for fault identification. It also carries out trend analysis and predicts when threshold limits will be exceeded.

(viii) The expert system implemented at Kori-2 (Korea Electric Power Corporation) PWR nuclear power plant diagnoses problems in the reactor coolant pumps, the control rod system, and the pressuriser. The reactor coolant pump is divided into the motor, seal, and hydraulic systems domains. When an anomalous event arises, the expert system tries to determine a primary causal alarm (obvious symptom) from the multiple alarms. The instrument readings, parameter trends, etc., are the non-obvious symptoms. Both symptoms are analysed to assist the operator in taking emergency and follow-up actions. This expert system differs from the other systems discussed in this section in that it is designed to assist the operating staff, and not the maintenance staff.

6.5 SUMMARY

The instrumentation required in sodium systems is of a specialised nature. The basic philosophy is to monitor process parameters while minimising physical intervention in the system that can result in a potential leak, and this is done primarily by exploiting the electrical and magnetic properties of the liquid metal coolant. The chapter closes with a brief discussion of the online conditioning monitoring systems used in light water reactors to highlight the advantages of implementing systems of a similar nature in fast reactors.

REFERENCES

1. O.J. Faust (Ed.), *Sodium-Na–K Engineering Handbook*, vol. III, Sodium Systems, Safety, Handling and Instrumentation, Gordon and Breach, Science Publishers, Inc., 1978.
2. G.E. Turner, Liquid Metal Flow Measurement (Sodium) State of the Art Study, Report no. LMEC-Memo-68-9, Liquid Metal Energy Center, 1968.
3. W.R. Miller, High Temperature Pressure Transmitter Evaluation, Report no. ORNL-2483, Oak Ridge National Laboratory, 1958.

4. Clyde C. Scott, Fermi Process Instrumentation, *Proceedings of the Symposium on Liquid-Metal Instrumentation and Control*, Idaho Falls, Idaho, March 1–2, 1967.
5. G.E. Turner, SCTI Sodium Instrument Operating Experience, *Proceedings of the Symposium on Liquid-Metal Instrumentation and Control*, Idaho Falls, Idaho, March1–2, 1967
6. R.H. Greene, D.A. Casada, C.W. Ayers, C.C. Southmayd, Detection of pump degradation, NUREG/CR-6089 ORNL-6765, Oak Ridge National Laboratory, August 1995.

Chapter 7

Operating experience of reactor centrifugal sodium pumps

7.1 INTRODUCTION

This chapter discusses the operating experience of main coolant centrifugal pumps in sodium-cooled reactors. The focus of the chapter is to highlight the operating conditions, the technical problems that have cropped up during commissioning and operation, and the solutions adopted. Hence only those reactors that have experienced problems during operation are the subject of interest.

The discussion will assist in highlighting the critical areas that deserve serious consideration during the design, manufacture, assembly, and operation of sodium centrifugal pumps.

7.2 OPERATING EXPERIENCE OF PUMPS

7.2.1 Experimental reactors

7.2.1.1 Experimental breeder reactor (EBR-II) [1–3]

The primary circuit of EBR-II consists of two centrifugal sodium pumps that operate in parallel to circulate coolant through the core. A single electromagnetic pump circulates sodium in the secondary circuit.

The experience with the primary centrifugal sodium pumps is described below.

Pump 1 seized after only a few hours of operation, and it was removed, cleaned of sodium, and examined. Severe galling of the lower labyrinth seal and the pump shaft was observed. The shaft was bowed by 0.05" (1.3 mm), and the thermal radiation baffle plates surrounding the shaft (in the cover gas space) were off-centre by 0.075" (1.9 mm) and in contact with the shaft.

Investigations revealed that bowing of the shaft resulted from incomplete stress relieving, which was done at 700°F (371°C) instead of at 900°F (482°C). It was also seen that the design of the labyrinth in the reactor pump was different from that in the prototype pump (which had been extensively tested for 16,000 h without any failure). In the prototype pump, the

DOI: 10.1201/9781003460350-7

labyrinth seal had thin teeth with deep grooves between the teeth, which caused rapid erosion of the teeth without damage to the shaft. In the reactor pump, however, the teeth were thicker with shallow grooves between them, as a result of which any contact between the shaft and the labyrinth resulted in damage to the shaft that was further aggravated by the chips in the shallow grooves resulting in severe galling of the shaft.

The problem was resolved by replacing the damaged shaft with a new shaft that was properly stress relieved. The labyrinth was also replaced, and the radial clearance between the shaft and labyrinth was increased from 0.015" (0.4 mm) to 0.124" (3.1 mm). The clearance between the shaft and the bottom thermal baffle was increased from 0.120" to 0.187" (3–4.7 mm) so that the clearance between the shaft and all four thermal baffles was uniform.

During the repair of Pump 1, Pump 2 also seized after ~200 h of operation. Similar repairs were done in Pump 2 also.

Following the above repairs, three incidents of minor binding of Pump 1 occurred, usually during periods following a long shutdown. These were attributed to sodium/sodium oxide accumulation in the radial clearance between the shaft and the lower labyrinth seal. Manual rotation of the shaft with the application of ~200 lb-ft (271.2 N-m) of torque was found to be sufficient to free the shaft.

Argon purge through the region was increased from 1 ft³/h (0.028 m³/h) to 5–7 ft³/h (0.14 m³/h–0.20 m³/h) to minimise the migration of sodium vapour up the shaft.

7.2.1.2 Enrico fermi atomic power plant (EFAPP aka Fermi) [4–7]

7.2.1.2.1 Primary pumps

During operation, overheating and noise were experienced with the thrust bearing of the rotating assembly, necessitating the replacement of the same with a new 200 lb preloaded bearing to reduce end play. The mechanical seals sealing the cover gas space required replacement (both stationary and rotary faces) in all three primary pumps.

The Fermi primary pump (refer Figure 2.5) differed from other sodium pumps in that two hydrostatic bearings supported the rotating assembly. The pump had alignment problems because of the rigid coupling, which was rectified by increasing the clearance of the upper bearing. This exercise reduced the number of effective bearings between the impeller and the lower motor bearing to only one (instead of the usual practice of having two bearings). The resulting frequent replacements of the mechanical seals was possibly due to this issue. High shaft torque was experienced for a period of two months, but this did not hinder pump operation.

The hydrostatic bearings (HSB) in Pumps 1 and 2 developed shallow scoring that required polishing. The journals of both upper and lower hydrostatic bearings required polishing in Pump 2 while the journal of the lower HSB required polishing in Pump 1. The bush of the lower HSB in Pump 1 was also cracked, and this was repaired by grinding the hardfacing.

Each pump was provided with a check valve at its outlet to prevent reverse flow through the pump during pump shutdown. The valves were of swing, disc type, and attached to the pump internals to facilitate removal of the valve along with the pump internals. The performance of the valves was satisfactory concerning the pressure drop across the valves. However, severe sodium hammer was experienced during their closure under reverse flow. The pumps were therefore removed and replaced with check valves of a different design that incorporated three antihammer devices: (i) spring loaded disc which stops the disc 12° from closure; (ii) re-designed body that permits more flow during the closure; and (iii) dashpot which reduces the closure force on the seal. The internals of the primary pumps were found to be in excellent condition when they were removed from sodium, after 7,000 h of operation (three years in sodium), to replace the check valves.

7.2.1.2.2 Secondary pumps

The secondary pump had only one hydrostatic bearing, viz.., in sodium. The upper end of the secondary pump shaft had two seals and the rotor weight was supported by a combination of radial and thrust bearings. As in the case of the primary pumps, the mechanical seals were frequently replaced. The eddy current coupling on one of the pumps was also repaired.

The three primary pumps operated for approximately 26,000 h, 27,000 h, and 28,500 h, respectively, at temperatures ranging from 400°F (204°C)–750°F (399°C) and flow rates up to 11,800 gpm (0.74 m³/s), while the three secondary pumps operated for approximately 17,000 h, 19,000 h, and 23,000 h respectively at temperatures in the range400°F (204°C)–750°F (399°C), and flow rates up to 13,000 gpm (0.82 m³/s).

7.2.1.3 Hallam nuclear power facility (HNPF) [3, 6, 7]

7.2.1.3.1 Primary pump

The sodium level in the pump tank dropped below the normal operating level when the pump flow rate was increased beyond 60% of the design flow rate, causing the entrainment of argon gas into the impeller. The drop in the sodium level was due to comparatively smaller leakage into the pump tank cover gas space (through the seal between the high-pressure bowl and lower pressure cover gas space and from the hydrostatic bearing) than that out of it (through the overflow line and through the balancing holes in the impeller). The problem was overcome by reducing the outflow from the pump tank cover gas region by plugging four of the eight balancing holes in the impeller.

The primary pumps operated without any mechanical problems except for an intermittent squeal. Disassembly and examination of one pump revealed no mechanical issues. The squeal was therefore attributed to noise from resonance in the piping system.

Pump operation was smooth, and there were no problems until welding repair work was undertaken on one of the sodium heat exchangers. Within a short time after this repair, all three primary pumps failed. Disassembly and examination showed that foreign material had lodged in the wearing ring clearances, which could have entered the system during the welding repair. A diagonal wear pattern on the hydrostatic bearing was attributed to misalignment and rubbing of the bearing due to uneven temperature distribution around the circumference of the pump casing at the bearing elevation. The thermocouple temperature readings recorded during the operation confirmed the uneven temperature distribution. It was supposed to have resulted from the external fins provided on the pump casing for forced cooling while cooling in the plant was by natural convection during pump operation. Moreover, the axial flow of sodium between the pump shaft and casing, envisaged in the design, was also partially blocked.

7.2.1.3.2 Secondary pump

Binding of the rotating assembly was experienced due to wearing out in close clearances such as impeller suction and discharge wearing rings and hydrostatic bearing. The binding was attributed to foreign particles in these spaces compounded by thermal distortion of the pump casing. The problem was resolved by filtering the sodium in a bypass loop, providing forced cooling of the pump casing, and increasing the running clearances of both the suction and discharge wearing rings.

7.2.1.4 Rapsodie [7–9]

Difficulties with the hydrostatic bearings of both primary and secondary pumps were experienced. During the isothermal commissioning tests, one primary pump and one secondary pump seized. The clearance of the hydrostatic bearing was increased from 0.280 mm to 0.400 mm, and the length-to-diameter ratio of the bearing was also changed. In the case of the secondary sodium pump, the hardfacing of the bush and journal was changed from nickel-based Colmonoy to cobalt-based Stellite. Other problems experienced with the primary pump were a malfunction of the check valve at the pump discharge and sodium oxide contamination. Problems related to vortexing and gas entrainment in the primary pump were resolved through simulated water tests.

7.2.1.5 Bystrij Opytnyj Reactor (BOR-60) [10]

One primary pump experienced high vibrations necessitating replacement. Inspection showed that the shaft was deformed by 1 mm, possibly due to incorrect heat treatment after welding composite parts of the shaft assembly. No failures were detected in the other pumps.

7.2.1.6 JOYO [7, 11–13]

Primary Pump P-1A was replaced in 1983 because of distortion of the pump's inner casing. This resulted from circumferential temperature variation from 323 K to 277 K due to convection currents in the cover gas space between the inner and outer casings. The pump was replaced with a modified one incorporating baffles in the cover gas space between the inner and outer casings to impede the convection currents. This modification resulted in a reduction of the circumferential temperature difference from 50°C to 4°C. Minor issues were also experienced with the mechanical seal lubrication and the coupling of the secondary sodium pump.

7.2.1.7 Kompakte Natriumgekuhlte Kernreaktoranlage (KNK II) [14, 15]

The primary sodium pump in Circuit 2 showed increase in vibrations (from 7–0 µm to 250 µm) after several years of trouble-free operation.

The pump had operated trouble-free for about 90,000 h (65,000 h at 200°C and 25,000 h at 520°C) when the increase in vibrations was detected. Examination revealed that the shaft had bowed, possibly due to annealing out of residual stresses or magnification of an initial unbalance due to thermal creep.

The primary pumps have operated for a cumulative period of more than 150,000 h while the secondary pumps have operated for a cumulative period of more than 110,000 h.

7.2.1.8 Fast flux test facility (FFTF) [16–18]

Shortly after the initial criticality and prior to raising of power there was an accidental overfilling of the pump. The sodium level was raised approximately 1.8 m above the normal operating level permitted during pony motor operation. During this incident, the reactor was in shutdown mode, and the pump was not operational. The overfilling resulted in the sodium baffles becoming wet and sodium freezing on the rotor surface in the area above the thermal baffles. As a result, the rotor assembly became imbalanced. Additionally, the uneven thermal gradient in this region caused the shaft to deform. When the pump was restarted, the measured vibration was nearly 20 times higher than before the overfilling incident and the vibration amplitude at the operating speed of 1000 rpm was deemed unacceptable. After disassembling and cleaning the pump, runout measurements were conducted on the shaft, revealing a 1.7 mm bend in the region that was overfilled.

During a shutdown to replace the motor of Pump 1, Pump 3, which was operating on pony motor, seized. High torque was initially unsuccessful in releasing the seizure. Heating the top portion of the pump and applying torque, however, returned the pump to normalcy without requiring its removal from the pump tank.

Investigation revealed that the cause of the seizure was the relocation of a sodium compound deposit that had formed in the cool top portion of the pump shaft/thermal baffle annulus during the earlier flooding incident. Although this deposit had a significant influence on pump operation with the pony motor, there was no significant effect on pump operation with the main motor.

Heating of the upper portion of the pump during subsequent operation and periodic measurement of the clearance in the annulus between the shaft and the shield plug was proposed to prevent the recurrence of the incident.

7.2.1.9 Fast breeder test reactor (FBTR) [19, 20]

After approximately 10,000 hours of operation, a rattling noise emanated from the discharge of the secondary sodium pump in the west loop. Upon inspection, no signs of loose piping supports or fixtures were found. A radiographic examination was attempted to detect foreign material in the pump, but it proved ineffective. The pump was operated with the sodium level lowered below the anti-vortex plate to rule out the possibility of noise emanating from loose objects trapped in the anti-vortex plate of the pump. However, this exercise did not produce any change in the noise. The pump was then operated at various speeds and flow rates to investigate if the noise originated from flow-induced vibrations. It was observed that the noise persisted at all speeds above 330 rpm. The pump was also operated with the cover gas pressure increased to 200 mbar(g) to rule out the possibility of cavitation, but the noise persisted. Analysis of vibration readings revealed no significant increase compared to the baseline data. Acoustic emission measurements yielded inconclusive results. Finally, the pump was removed, cleaned, and inspected thoroughly, uncovering the rubbing marks on the shaft end plug and labyrinth. Consequently, the pump was replaced with a spare unit, which operated smoothly.

7.2.2 Demonstration reactors

7.2.2.1 Phenix [14, 15, 21, 22]

During commissioning tests before fuel loading, the non-return valves at the primary pump outlet failed to work. All the pumps were removed, and the valves were replaced.

Vibrations occurred in one primary pump due to the loosening of the hydrostatic bearing journal from the shaft during thermal transients. A similar failure occurred in the secondary sodium pump also. The design was modified, and replacements were made on all the pumps, including the spare pumps. Cracks were observed on the flange of the replaced non-return valve of one of the pumps.

This was attributed to caustic stress corrosion cracking resulting from imperfect drying when the valve was replaced earlier.

The maximum number of operating hours that a primary pump has completed is 220,000 h. The highest number of unscheduled shutdowns for a primary pump was 21, while that for a secondary pump was 24, due mainly to the failure of thyristors in the motor speed regulation system.

7.2.2.2 Prototype fast reactor (PFR) [14, 23–25]

The primary sodium pumps of this reactor are of double suction type. During the commissioning trials in 1973, the primary sodium pump seized during restarting after a shutdown of several hours. The pump stopped due to an overload failure of the shear pins between the pump and motor shafts. Disassembly and examination revealed that severe rubbing occurred between the shaft and the inner bore of the top shield plug, resulting in metal transfer between the parts. The shaft was also observed to be bowed. Seizure had also occurred at the bottom hydrostatic bearing. The filters upstream of the feed to the bearing were observed to be clean, and there was no evidence to suggest that the hydrostatic bearing was the initiator of the seizure. It was concluded that the bowing of the shaft, from temperature asymmetry around the shaft, resulted in the closure of the clearance between the shaft, and the inner bore of the shield plug. The localised frictional heating further compounded the bow, increasing the rubbing, local damage, and overloading of the bottom hydrostatic bearing.

During the commissioning exercise, the secondary pump seized at the hydrostatic bearing location, following a speed increase from 200 rpm to 930 rpm, causing the drive motor to trip on overload. The cause was the detachment of the stellite coating on the bearing journal. The investigation also revealed that acoustic sensors mounted on the pump casing had provided an early indication of rotor rub. The problem recurred in 1984. In 1987, the bottom labyrinth seal on the secondary sodium Pump 2 cracked, and the unit jammed due to excessive clearance in the top bearing requiring the replacement of the fluid coupling of the pump. In light of the experience of seizure with the primary and secondary pumps, it was decided to provide barring gear on all pumps to prevent shaft bending when the pump is stationary for an extended period.

It was also decided to provide additional instrumentation to monitor pump shaft behaviour and detect any abnormality at an early stage.

A major oil leak from the top bearing and seal of the primary sodium Pump 2 occurred in June 1991 during the venting of the argon gas blanket to rectify low flow rate in the argon gas blanket circulating system. During the venting exercise, sodium in the pump tank entered the bearing oil drain tank. This was due to blockage of the overflow pipe (connecting the pump tank free surface to the pump suction/cold pool), resulting in a pressure differential between the gas blankets in the main vessel and the pump casing. The venting process resulted in a flow of gas from the pump casing, thereby reducing the pressure there and causing sodium to rise up the shaft and into the bearing oil drain tank. The sodium displaced the oil in the tank into the sodium in the pump casing. About 17 litres of oil was displaced into the sodium in the process.

As early as 1994, an oil leak was suspected when PSP 2 was removed for modifications to its instrumentation, and black sooty deposits were noticed on its surface. It is suspected that this resulted in the failure of the filter of the check valve at the pump outlet and partial blockage of the overflow line resulting in the incident in 1991.

After the incident in 1991, all three valve and filter assemblies were replaced, leading to a shutdown of 18 months.

Examination of the removed parts showed that at least one panel of each valve filter had failed, and there was oil-related debris on all the filters.

The primary pumps had operated for a cumulative period of 405,965 h, while the secondary pumps had operated for a cumulative period of 231,960 h.

7.2.2.3 Bystrie Neytrony BN-350 [26–29]

Cavitation damage occurred in the region of the overflow orifices of the pump. Significant cavitation damage was observed in the impeller vanes of both the primary and secondary pumps. Hydraulic shock was experienced during the closing of the check valve at the outlet of the primary pump. The valve was replaced with an improved design. Some issues periodically experienced were: (i) an increase in torque during restarting of the pump after prolonged shutdown because of freezing of sodium in the narrow clearances between the shaft and stationary parts; (ii) fluctuations in sodium level in the pump tank, in Loop 4, due to variation in leakage at different temperatures; and (iii) increased vibration(>100 μm) of the pumps.

The level fluctuation in (ii) above was because of the arrangement provided in the BN 350 pumps to maintain the level in the pump tank. It may be recalled from Section 3.2.6.1 that the sodium level in the primary pump tank was maintained by allowing excess sodium to overflow into a drain tank. When the pump was operated at 1,000 rpm, self-induced vibration was experienced due to fluctuation in the sodium level in the drain tank. Investigation revealed that the slope of the pipeline connecting the pump tank to the drain tank was altered due to a change in the tension of the pipe spring hangers. This caused a temporary reduction in the sodium draining to the drain tank, causing the level in the drain tank to fall and that in the pump tank to rise. Once the level in the pump tank rose above a particular value, liquid draining from the pump tank increased, causing the drain tank level to rise and the pump tank level to fall.

It was observed that this fluctuation in level resulted in a self-induced vibration with a period of 1.6 s.

The problem was resolved by: (i) changing the spring tension of the pipe hangers connecting the pump tank to the drain tank; (ii) isolating the pump tank cover gas from the primary circuit cover gas; and (iii) maintaining the gas pressure in the tank higher than that in the primary circuit.

7.2.2.4 Bystrie Neytrony 600 (BN-600) [26, 27, 29, 30]

The BN-600 primary sodium pumps are of double suction type. During the course of attaining full power operation, increased vibrations were recorded at the top-bearing location. On inspection, it was observed that some teeth and springs of the coupling had broken. These were re-designed and replaced. During subsequent operation, high vibrations were experienced in the speed range of 925–940 rpm. It was found that the frequency-controlled motor coupled to the synchronous rectifier drive induced torsional fluctuation of 6%–8% of the nominal torque. Resonance occurred when the torque fluctuation coincided with the natural frequency of the rotating assembly, resulting in cracks in the shaft and failure of the coupling. Sometime in 1982–1983, strain measurements were made based on which shaft stresses were estimated, and a forbidden speed range identified. Later, the problem was entirely resolved by replacing the damaged shafts with modified ones of the same strength and changing the pump control algorithm. In the revised control scheme, the pump frequency control was suspended by switching off the synchronous rectifier stage after a prescribed power level was reached.

The pump also experienced cavitation damage which came to light when the pump was taken for maintenance after about 40,000–50,000 h of operation. The impeller of the primary pump in BN-600 was similar in design to that in the BN 350 primary pump. Therefore the cavitation damage observed on the BN-600 primary pump was similar to that in the BN 350 primary pump. The damage was observed on the suction face of the vanes near the inlet edge and in some places on the impeller shroud. The dimensions of the damaged area was about 150 mm in length, 70 mm in breadth, and 18 mm in depth. The damage seen is a case of classical cavitation due to the low cover gas pressure in the main vessel and the prolonged operation in non-optimal mode. It was observed that the margin of 1.8 on the NPSH requirement needed to be increased, and the desired value was about 4. Although the wear of the impeller vanes did not influence the operating parameters of the pump, the resulting dispersion of wear debris into the primary sodium was not desirable.

7.3 SUMMARY

Centrifugal Sodium Pumps have operated successfully as main coolant pumps of sodium-cooled reactors. Although problems have been experienced in the early stages of development, these are only to be expected in the natural course of the development of this complex equipment, and it is encouraging that the problems have been successfully overcome.

It is observed that a recurring problem in many pumps (EBR-II, HNPF, PFR) was binding/seizure of the rotating assembly in the labyrinth region in the cover gas space. These events have emphasised the importance of tailoring equipment design to address issues specific to the operating environment and temperature, such as sodium system-specific problems like the deposition of sodium vapour in cooler regions in cover gas space; the significance of straightness of slender parts (e.g., shaft), closure of the clearance between parts operating with temperature differential as well as the importance of proper heat treatment/manufacturing procedures.

A study [31] using data obtained from Centralized Reliability Data Organization (CREDO) on EBR-II, FFTF, and JOYO pumps has shown the cumulative event rate for all pumps as 34.4 event/million operating hours with 5% and 95% one-sided confidence limits of 26.3 and 44.4 event/million operating hours respectively.

Specifically, the cumulative event rates for EBR-II, FFTF, and JOYO were 30.0, 32.4, and 46.4 event/million operating hours, respectively.

REFERENCES

1. J.R. Davis, G.E. Deegan, J.D. Leman and W.H. Perry, Operating Experience With Sodium Pumps at EBR-II, Report no. ANL/EBR-027, October 1970.
2. B.C. Cerutti, G.E. Deegan, J.D. Nulton, W.H. Perry and R.E. Seever, Removal and Repair of EBR-II Primary Sodium Pump No. 1, Report no. ANL-7835, May 1972.
3. Sodium Pump Development and Pump Test Facility Design, Report no. WCAP-2347, Westinghouse Electric Corporation, August 1963.
4. D.J. Kniley, W.J. Carlson, E. Ferguson, and O.G. Jenkins, Mechanical Elements Operating in Sodium and other Alkali Metals, Vol. II, Experience Survey, LMEC-68-5, June 1970.
5. J.G. Duffy, and H.A. Wagner, Operating Experience with Major Components of the Enrico Fermi Atomic Power Plant, *Proceedings of a Symposium on Performance of Nuclear Power Reactor Components*, Prague, 10–14, November 1969.
6. W. Babcock, State of Technology Study-Pumps: Experience with High Temperature Sodium pumps in Nuclear Reactor Service and their Application to FFTF, BNWL-1049, December 1969.
7. P.G. Smith, Experience With High Temperature Centrifugal Pumps in Nuclear Reactors and their Application to Molten Salt Thermal Breeder Reactors, ORNL-TM-1993, September 1967.

8. L. Vautrey, Fast Reactor Development in France, *IWGFR First Annual Meeting*, Vienna, March 1968.
9. L. Vautrey, The Development of Fast Neutron Reactors in France from March 1969 to March 1970, *IWGFR Third Annual Meeting*, Cadarache, March 1970.
10. Operational and Decommissioning Experience with Fast Reactors, Proceedings of a Technical Meeting held in Cadarache, Report no. IAEA-TECDOC-1405, France, 2002.
11. F. Asakura, The Experience of Experimental Fast Reactor JOYO Operation and Maintenance, *International Conference on Fast Reactors and Related Fuel Cycles*, Vol. 1, November 1991, Kyoto, Japan.
12. Y. Matsuno, JOYO Operating Experience: 1977–1983, Liquid Metal Engineering Technology, *Proceedings. of the Third International Conference* held in Oxford on 9–13, April 1984.
13. Masao M. Hori, and Yoshihiko Y. Nara, Operating Experiences of Experimental Fast Reactor JOYO, *International Conference on Fast Reactors and Related Fuel Cycles*, Vol. 1, November 1991, Kyoto, Japan.
14. M. Sauvage, A.M. Broomfield, and W. Marth, *Overview on European Fast reactor Operating Experience, International Conference on Fast Reactors and Related Fuel Cycles*, Vol. 1, November 1991, Kyoto, Japan.
15. C.V. Gregory, and C. Acket, Repairs in Fast Breeder Reactors: Experience to Date, Prospects Ffor The Future, *International Conference on Fast Reactors and Related Fuel Cycles*, Vol. 1, November 1991, Kyoto, Japan.
16. Stuart A. Krieg, James D. Thomson, *Fast Flux Test Facility Replacement of a Sodium pump, HEDL-SA-3364-FP, DE 87 004092*, Westinghouse Hanford Company, Salk Lake City, UT, March 1986.
17. C.J. Peckinpaugh, R.A. Bennett, and W.R. Wykoff, Fast Flux Test Facility (FFTF) Operational Results, *Proceedings of International Conference on Nuclear Power Experience*, Vol. 5, September 1982.
18. Q.L. Baird and R.A. Harris, FFTF Operational Results: Startup to 1000 MWd/kg, *Nuclear Safety*, Vol. 26, No. 2, March–April 1985.
19. B. Rajendran, P.V. Ramalingam, T.R Ellappan, A.P. Chaba, R. Veerasamy, Amitava Sur & M.K. Ramamurthy, Experience Gained During the Commissioning of Sodium Systems of FBTR, *3rd National Symposium on Operating Experience of Nuclear Reactors and Power Plants*, Bombay, 15–17, March 1989.
20. B. Rajendran, S.K. Chande and S.B. Bhoje, Experience on Sodium Pumps in FBTR, Seminar on Movement of Fluids in Nuclear Industry, Indira Gandhi Centre for Atomic Research, Kalpakkam, India, November 7–8, 1990.
21. Liquid Metal Cooled Reactors: Experience in Design and Operation, IAEA-TECDOC-1569, December 2007.
22. Didier D. Dall'Ava, Laurent L. Martin, and Bernard B. Vray, 35 Years of Operating Experience of PHENIX NPP SodiumCooled Fast Reactor, *Proceedings of the 17th International Conference on Nuclear Engineering, ICONE 17*, July, 12–16 2009, Brussels, Belgium.
23. A.D. Evans, J.M. Laithwaite, and D.A. Nunn, Design, Construction and Commissioning of PFR as at 14, December 1973, *Proceedings of the conference on Fast Reactor Power Stations*, London, March 1974.
24. A. Cruickshank, and A.M. Judd, Problems Experienced During Operation of the Prototype Fast Reactor, Dounreay, 1974-1994, IAEA TCM on Unusual

Occurrences During LMFBR Operation: Review of Experience and Consequences for Reactor Systems, November 1998.

25. C.V. Gregory, A Review of the Operation of the Prototype Fast Reactor, *Nuclear Energy*, Vol. 31, no. 3, 1992, pp. 173–183.

26. S.A. Belov, F.M. Mitenkov, E.G. Novinski, G.M. Nikolushkin and G.P. Shishkin, Design and experimental development of Sodium Pumps, Paper no. 121, Liquid Metal Engineering Technology, Proc. of the *Third International Conference held in Oxford* on 9–13, April 1984.

27. Status of Liquid Metal Cooled Fast Reactor Technology, IAEA-TECDOC-1083, April 1999.

28. M.F. Troyanov, and A.A. Rinejskij, Status of Fast Reactor Activities in the Russian Federation, *25th Annual Meeting of IWGFR*, Vienna, 1992.

29. A.A. Rineisky, V.N. Chushkin, A.A. Kamaev, N.N. Oshanov, and O.A. Potapov, The Utility of BN-600 Reactor Operating Experience on the Choice of Design and Technological Decisions of Fast Reactors Under Design, Indo Soviet Seminar, Obninsk, March 1989.

30. https://issuu.com/johna.shanahan/docs/t9.1.potapov

31. Mark S. Smith, Darren H. Wood, and James D. Drischler, An Assessment of Liquid-Metal Centrifugal Pumps at Three fast Reactors, *Nuclear Technology*, Vol. 104, No. 1, Oct. 1993, pp. 118–127.

Chapter 8

Cavitation in centrifugal pumps

8.1 CAVITATION

Cavitation occurs in a flowing liquid when the static pressure in the flow field falls below the saturated vapour pressure of the liquid at the operating temperature. When the local static pressure falls below the vapour pressure, the weakness of the liquid to resist tensile forces results in the vaporisation of the liquid and the adherence of the resulting vapour bubbles and stable vapour cavities on the containing walls produces disturbances in the flow field. The flowing liquid carries these bubbles away, and as soon as they enter a zone where the static pressure exceeds the saturated vapour pressure, the bubbles collapse (condense).

The implosion of a vapour bubble occurs rapidly, resulting in an onrush of the surrounding liquid into the imploding space. The asymmetrical collapse of the bubble results in a high-speed liquid microjet that impinges on the surface containing the flowing medium, producing surface damage. In addition, the rebounding bubble, from the pressure of its contents, produces shock waves capable of causing further damage. The pressure pulses produced during cavitation have been estimated to be of the order of thousands of atmospheres [1]. The repeated impingement of the high-speed water jets/ shock waves results in fatigue failure of the metal surface, often accelerated by thermal/electrical/chemical action. This phenomenon is known as cavitation damage or cavitation erosion. However, much before cavitation damage, the formation, growth, and collapse of vapour bubbles produce flow rate fluctuations accompanied by noise and vibration.

8.2 DEVELOPMENT OF CAVITATION IN PUMPS

As liquid flows through the suction approach of a pump and enters the pump inlet, there is a continuous reduction in static pressure.

Further reduction in pressure occurs as the liquid manoeuvres through the pump suction casing and enters the impeller flow passage into the front and back of the blades. The liquid entering the front of the blades suffers an

DOI: 10.1201/9781003460350-8

additional pressure drop before energy addition begins in the impeller. This difference between the pressure at the entry to the impeller blade and the minimum pressure on the blade surface is known as 'dynamic depression head'. The dynamic depression head is a function of the pump speed, flow velocity, pump head, and the specific speed of the pump [2]. Cavitation occurs when the lowest pressure in the flow path falls below the saturated vapour pressure of the liquid at the prevailing/operating temperature.

There are several stages in the development of cavitation in a pump. Consider a pump operating at constant speed and flow rate/capacity. As the NPSH is reduced, cavitation begins even at large values of available NPSH. Appreciable cavitation bubble activity is present at values of NPSH much higher than $NPSHR_{3\%}$, which is the parameter conventionally used to detect pump cavitation in the field. However, this does not affect the pump performance (flow rate and head) or lead to damage. This is referred to as cavitation inception and is detected either visually (in the case of water pumps) or, more generally, using acoustic techniques. Additional reduction in NPSH produces changes in the pump performance parameters, e.g., fluctuation in head and flow rate, and possible erosion over an extended period. It is seen from test results collected over 15 years from various US and European laboratories [3] that NPSH value as high as four times the $NPSHR_{3\%}$ may be required to ensure damage-free operation of the pump and as high as ten times the $NPSHR_{3\%}$ may be required to eliminate cavitation bubble noise. Continued reduction in NPSH results in a sharp decrease in the pump head, power, and efficiency for a given speed and capacity.

Figure 8.1 shows the development of cavitation, with NPSH reduction, in a pump. In the figure $NPSH_{incipient\ visual}$ is the NPSH value at which bubbles are first visible to the naked eye (around few millimeters in size) in water

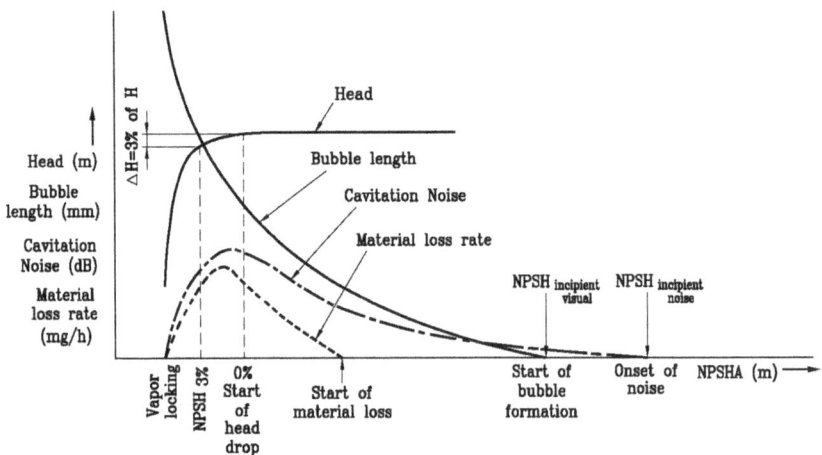

Figure 8.1 Variation of cavitation bubble size, noise, damage rate, and pump head with NPSH.

tests under stroboscopic lighting; $NPSH_{incipient\ acoustic}$ is the NPSH value at which high-frequency noise from cavitation bubble collapse is detected. The noise and erosion damage increase with a reduction in NPSH, reach a peak and then decrease as (qualitatively) shown in Figure 8.1. In pumps, cavitation noise measurement is easier to measure, requires no intervention in the system and is, therefore, preferred when compared to measurement of erosion damage; it is used to define the acceptable cavitation intensity in power plant pumps. There is a debate on whether acoustic noise is an earlier indication of cavitation inception or bubble formation precedes it. In an opinion survey conducted at the specialists' meeting on "Cavitation criteria for designing mechanisms working in sodium application to pumps", organised by IWGFR [4], on the order in which the thresholds of acoustic noise, bubble formation, erosion damage, and 3% head loss occur, in a test involving a gradual reduction of suction pressure to induce cavitation, the participants agreed that the acoustic threshold was most sensitive to cavitation onset and therefore the first to be detected.

8.3 CONDITIONS FOR CAVITATION IN PUMPS AND EXPRESSION FOR NET POSITIVE SUCTION HEAD REQUIRED (NPSHR)

For normal working of a pump without cavitation, it is essential that the minimum value of static pressure in the pump, i.e., in the front face or suction face of the blade, P_{min} is greater than the saturated vapour pressure, P_v of the pumped liquid, i.e.

$$P_{min} > P_v$$

Consider a liquid flowing through the pump intake at velocity V_{in} at pressure P_{in}. The Net Positive Suction Head (NPSH) of the liquid at the pump intake is given by the expression:

$$NPSH = \frac{P_{in}}{\gamma} + \frac{V_{in}^2}{2g} - \frac{P_{vp}}{\gamma}$$

where γ is the specific weight of the liquid.

Figure 8.2 shows the suction and pressure surfaces of the blade on the end view of the impeller.

As the liquid approaches the impeller vanes there is further drop in pressure due to losses from viscous friction, acceleration of the liquid and shock loss at the blade entry. These losses are given by the relation:

$$\frac{C_1^2}{2g} + \lambda \frac{W_1^2}{2g} \tag{8.1}$$

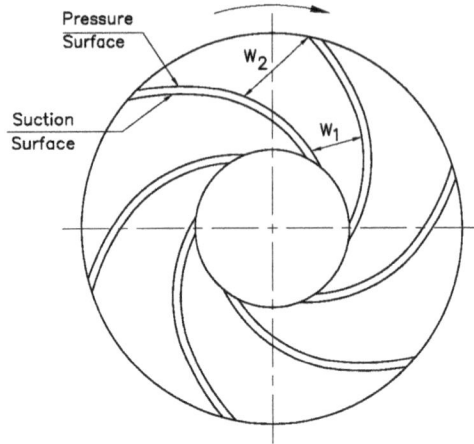

Figure 8.2 Suction and pressure faces on pump impeller.

where C_1 is the absolute velocity of liquid entering the impeller blade, W_1 is the relative velocity of liquid entering the impeller blade, and λ is a constant that depends on the blade profile and flow conditions.

At the design flow rate there is no prerotation, hence:

$$C_1 = C_{m1} = \frac{Q}{\pi \frac{D_1^2}{4}}$$

In the actual pump, the meridional flow velocity at the front shroud inlet where cavitation begins is higher than that calculated from the flow rate divided by the eye area because of (i) the curvature of the flow; and (ii) the reduction in the net flow area at blade inlet due to the thickness of the vanes.

Cavitation will occur if:

$$\frac{P_{in}}{\gamma} + \frac{V_{in}^2}{2g} - \left\{ \frac{C_1^2}{2g} + \lambda \frac{W_1^2}{2g} \right\} \le \frac{P_{vp}}{\gamma} \tag{8.2}$$

In order to avoid cavitation:

$$\frac{P_{in}}{\gamma} + \frac{V_{in}^2}{2g} - \frac{P_{vp}}{\gamma} \ge \frac{C_1^2}{2g} + \lambda \frac{W_1^2}{2g}$$

From the velocity triangle at the pump inlet (for no prerotation):

$$W_1^2 = C_1^2 + U_1^2$$

Therefore:

$$\frac{P_{in}}{\gamma} + \frac{V_{in}^2}{2g} - \frac{P_{vp}}{\gamma} \geq (1+\lambda)\frac{C_1^2}{2g} + \lambda\frac{U_1^2}{2g}$$

In the limiting case, the left-hand side of the above equation becomes equal to the right-hand side and this corresponds to the beginning of cavitation. Therefore:

$$\text{NPSHR} = (1+\lambda)\frac{C_1^2}{2g} + \lambda\frac{U_1^2}{2g} \tag{8.3}$$

Thus the Net Positive Suction Head Required (NPSHR) of the impeller depends on C_1 and U_1 which in turn are dependent on the pump flow rate, Q, pump speed, N, and the eye diameter, D_1. Brennen [5] quoting the experimental work of Gongwer [6] provides empirical correlations for cavitation free and break down NPSH as below.

$$\text{Cavitation free NPSH} = 1.8\frac{C_1^2}{2g} + 0.23\frac{U_1^2}{2g} \tag{8.4}$$

$$\text{Breakdown NPSH} = 1.49\frac{C_1^2}{2g} + 0.085\frac{U_1^2}{2g} \tag{8.5}$$

where C_1 and U_1 are in ft/sec.

Based on experiments/test data, SS Rudnev suggested an equation for determining NPSHR, as below [7]:

$$\text{NPSHR} = 10\left(\frac{N\sqrt{Q}}{C}\right)^{\frac{4}{3}} \tag{8.6}$$

The coefficient 'C' is experimentally determined from test data on different pumps and is a measure of the maximum permissible operating capacity of the pump without cavitation. The above equation can be re-written as

$$C = 10^{\frac{3}{4}}\frac{N\sqrt{Q}}{\text{NPSHR}^{\frac{3}{4}}} = 5.62\frac{N\sqrt{Q}}{\text{NPSHR}^{\frac{3}{4}}} \tag{8.7}$$

It is clear from the above form of the equation that C is representative of the suction specific speed of the pump.

8.4 THERMODYNAMIC EFFECT ON PUMP CAVITATION

Although pure liquids have good tensile strength [8], the presence of microscopic free gas nuclei on the walls of the container results, in practice, in the vaporisation of the liquid when the static pressure falls below the saturated vapour pressure.

These nuclei grow either by the inward diffusion of gas from the liquid or by the rapid vaporisation of the surrounding liquid depending upon whether enough time is available, for the former case, or the local pressure is sufficiently below the saturated vapour pressure (for the latter case). Compared with water (the most common heat transfer medium), the thermal properties of liquid sodium are vastly different, and so are the growth, population, and collapse intensity of the vapour bubbles.

Reference [9] gives a detailed summary of the various liquid properties affecting cavitation and the damage produced from cavitation. This section discusses the 'thermodynamic effect' on cavitation in centrifugal pumps [10–12]. This concept is relevant to sodium pumps because the vaporisation of liquid during cavitation is essentially a thermal process, and the critical thermal properties of liquid sodium and water, the liquid commonly used to evaluate the hydraulic performance of pumps, including sodium pumps, are widely different.

Thermodynamic effect refers to the thermal restraint on bubble growth and collapse resulting from a change in the vapour (bubble) temperature due to the heat transfer of the bubble contents with the surrounding liquid. As a result, the latent heat of vaporisation or condensation is not transferred fast enough between the liquid and the bubble, thus impeding the growth/collapse of the bubble and reducing damage. The thermodynamic criterion was introduced by Stahl and Stepanoff [10–12] to explain the reduction in Net Positive Suction Head (NPSH) requirement for pumps handling hydrocarbons when compared to those handling water. The criterion expresses the ratio of the volume of vapour formed per unit quantity of liquid passing through the low-pressure zone for a unit reduction in pressure head under thermal equilibrium conditions. This effect is also responsible for the variation of cavitation behaviour, i.e., head drop, erosion rate, etc., with temperature in pumps.

Stepanoff expressed it using the relation:

$$B = \frac{V_V}{V_L} = \left(\frac{v_V}{v_L}\right) \times \left(\frac{\Delta h_f}{L}\right) \tag{8.8}$$

where V_v = total volume of vapour produced, V_L = total volume of liquid passing through the low-pressure region, v_v = specific volume of vapour, v_L = specific volume of liquid, Δh_f = enthalpy increase corresponding to a reduction in pressure below saturation conditions, L = latent heat of vaporisation of the liquid. The equation, however, does not consider the bubble formation and bubble collapse rate; i.e., the heat transfer rate, which depends on the liquid's latent heat and thermal diffusivity and the equilibrium bubble size. The expression was modified by Florschuetz and Chao [13] to include the effect of heat transfer effect on bubble growth and collapse and the improved expression is:

$$B_{eff} = \left(\frac{\rho_L c_L \Delta T}{\rho_v L} \right)^2 \frac{K_L}{R_0} \left(\frac{\rho_L}{\Delta P} \right)^{1/2} \qquad (8.9)$$

Expressing the equation in terms of specific volumes of liquid and vapour:

$$B_{eff} = \left(\frac{v_v c_L \Delta T}{v_L L} \right)^2 \frac{K_L}{R_0} \left(\frac{\rho_L}{\Delta P} \right)^{1/2} \qquad (8.10)$$

where:
  K_L = thermal diffusivity of the liquid = $k_L/(\rho_L c_L)$, ft²/hr
  k_L = thermal conductivity of the liquid, BTU/hr-ft-°F
  c_L = specific heat of the liquid, BTU/lbm-°F
  ρ_L = density of liquid, lbm/ft³
  v_L = specific volume of liquid, ft³/lbm
  ρ_V = density of vapour, lbm/ft³
  v_V = specific volume of vapour, ft³/lbm
  R_0 = equilibrium radius of the bubble, ft
  ΔP = reduction in pressure causing cavitation, lbf/ft²
  ΔT = reduction in temperature in the liquid film due to vaporisation, °F
  L = latent heat of evaporation, BTU/lbm

The thermodynamic parameter B (Equation 8.8), which does not take into account the heat transfer effects, can be expressed in terms of the Jakob number,[1] Ja and the NPSH as [14]

  Ja = B.NPSH

where:

  Ja = Jakob number = $\left(\dfrac{v_v c_L \Delta T}{v_L L} \right)$

Substituting for Ja in (8.10) gives:

$$B_{eff} = B^2 \text{NPSH}^2 \frac{K_L}{R_0} \left(\frac{\rho_L}{\Delta P} \right)^{1/2} \qquad (8.11)$$

Expressing ΔP in equation (8.11) in terms of NPSH gives:

$$B_{eff} = B^2 \frac{K_L}{R_0} \left(\text{NPSH} \right)^{3/2} \qquad (8.12)$$

Since the thermal conductivity of liquid sodium is higher than that of water, and the volumetric heat capacities of the liquids are similar, the vaporisation produced in liquid sodium by a local pressure drop is much more vigorous than that in water resulting in a large volume of vapour per volume of liquid passing through the cavitation zone (i.e., large value of B_{eff}). The growth and collapse of such large vapour volumes are inertia controlled, resulting in high jet velocities and increased damage. On the contrary, for small values of B_{eff} (typically < 1,000 for liquid metals), the bubble growth and collapse are governed by heat transfer effects [15].

Table 8.1 [14, 15] calculated using equation (8.10) above for nominal values of bubble radius, R_0 (1 cm) and NPSH (1 ft) compares the values of B_{eff} in water and sodium. It is seen that in the case of sodium, the thermodynamic effect becomes important at temperatures above 1,500°F (816°C).

Pump head is reduced at low NPSH values because the energy added to the fluid by the pump is largely absorbed by the vapour passing through the impeller instead of being transferred to the liquid. Large vapour volume at the inlet (as indicated by large value of B_{eff}) will produce a bigger reduction in the pump head.

It is seen from Table 8.1 that the volume of vapour produced decreases substantially with increasing temperature for both water and sodium. This

Table 8.1 Variation of B_{eff} with temperature for water and sodium

Ser	Liquid and temperature	B_{eff}
1	Water (55°F/12.8°C)	174
2	Water (95°F/35°C)	1.35
3	Water (120°F/48.9°C)	0.10
4	Water (150°F/65.6°C)	0.006
5	Water (180°F/82.2°C)	0.0005
6	Sodium (500°F/260°C)	1.12×10^{15}
7	Sodium (1500°F/816°C)	0.01

reduction in vapour volume with increasing temperature is the reason for conducting pump cavitation tests in water at a controlled temperature close to room temperature where B_{eff} is large, and the effect of cavitation on the head drop is significant. Another pertinent observation from Table 8.1 is the significant value of B_{eff} for sodium at 260°C compared to that for water at room temperature, implying the volume of vapour produced in a pump cavitating in sodium is much more than that in a similar pump in water, and so the head drop and damage produced in sodium is more substantial than that in water.

The above considerations indicate that the damage produced during vapour bubble collapse is complex. It is observed experimentally that the cavitation damage in sodium, at the operating temperature, is more than that in water at room temperature. Reiser [16] has reported that the damage in sodium at 204°C is about 1.5 times that in water at 27°C.

8.5 TYPES OF CAVITATION AND DAMAGE LOCATIONS IN A PUMP

Table 8.2 summarises the various types of cavitation experienced in a pump and the damage locations.

Figure 8.3 shows the phenomena of suction and discharge recirculation in the impeller.

Figure 8.4 shows the damage locations on the impeller resulting from classical cavitation and recirculation.

Table 8.2 Damage from cavitation in pumps

Type of cavitation	Cause	Damage location
Classical	NPSHA < NPSHR$_{3\%}$ at the operating flow rate.	On blade suction surface slightly downstream of blade inlet.
Discharge recirculation	Operation at flow rate, Q_{DR} less than best efficiency flow rate, i.e., $(Q_{DR} < Q_{BEP})$.	On blade pressure surface close to impeller outlet. Can also result in damage to volute tongue, diffuser vanes, impeller shroud at outlet.
Suction recirculation	Operation at flow rate, Q_{SR} less than best efficiency flow rate, i.e., $(Q_{SR} < Q_{DR} < Q_{BEP})$.	On blade pressure surface close to impeller inlet. Can also result in damage to suction guide vanes.

Discharge
Recirculation

Suction
Recirculation

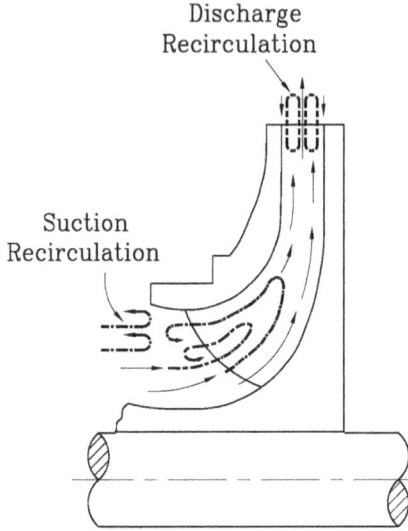

Figure 8.3 Suction and discharge recirculation.

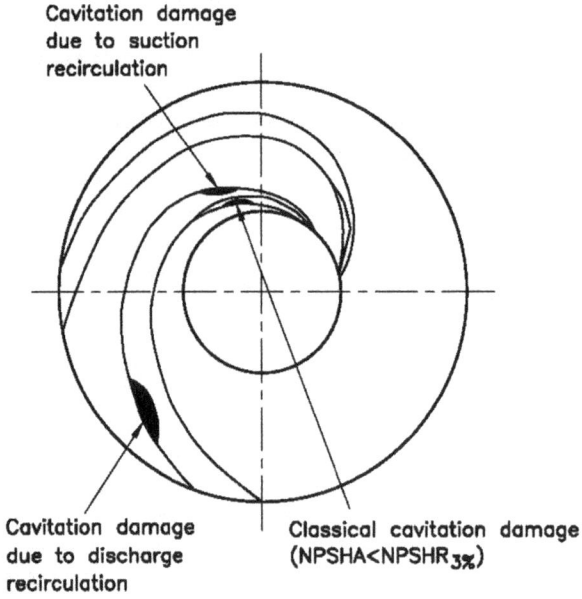

Cavitation damage
due to suction
recirculation

Cavitation damage
due to discharge
recirculation

Classical cavitation damage
(NPSHA<NPSHR$_{3\%}$)

Figure 8.4 Damage locations on the impeller due to classical cavitation and recirculation.

8.6 DESIGNING IMPELLER FOR IMPROVED CAVITATION PERFORMANCE [17, 18]

The following points are to be considered for designing impeller for improved cavitation performance:

1. The eye diameter for optimum cavitation performance can be calculated as given below.
   The critical NPSH given by Equation (8.3) above is:

$$NPSHR = (1 + \lambda)\frac{C_1^2}{2g} + \lambda\frac{U_1^2}{2g}$$

Expressing the velocities at pump inlet in terms of the eye diameter and pump flow rate:

$$NPSHR = \frac{1}{2g}\left\{(1 + \lambda)\frac{16Q_1^2}{\pi^2 D_1^4} + \lambda\frac{\pi^2 D_1^2 N^2}{60^2}\right\} \qquad (8.13)$$

where N is the operating speed of the pump in rpm.
   Differentiating w.r.t the eye diameter, $D1$, and equating to zero gives the optimum eye diameter for minimum NPSHR as:

$$\frac{d}{dD_1}(NPSHR) = -4(1 + \lambda)\frac{16Q_1^2}{\pi^2 D_1^5} + \frac{2\lambda D_1 \pi^2 N^2}{60^2} = 0$$

$$D_{1\,optimum} = 3.25 \sqrt[6]{\frac{(1 + \lambda)}{\lambda}}\sqrt[3]{\frac{Q}{N}} = k_1\sqrt[3]{\frac{Q}{N}} \qquad (8.14)$$

Rudnev has indicated that the value of k_1 is in the range of 4.3–4.5 depending on the value of λ (which, as mentioned earlier, depends on the blade profile and flow conditions). Turton [19] has indicated a value of 4.66.
   Substituting the expression for the optimum value of D_1 from (8.14) in (8.13) gives:

$$NPSHR_{min} = \frac{1}{20}\sqrt[3]{\frac{\pi^2}{15}}\sqrt[3]{\lambda^2(1 + \lambda)}\frac{Q^{\frac{2}{3}}N^{\frac{4}{3}}}{2g} = k_2\frac{Q^{\frac{2}{3}}N^{\frac{4}{3}}}{2g}$$

where k_2 is a factor that is a function of λ.
   In the design process, the flow rate, Q, and NPSH are usually specified, and the impeller diameter and speed are to be calculated.

Following a similar approach, Pearsall [20] gives expressions for the impeller diameter and operating speed for optimum cavitation performance as:

$$\text{Impeller diameter, } D_{t1}{}^2 = \frac{4Q}{\pi\left\{1-\left(\dfrac{D_h}{D_t}\right)^2\right\}}\left\{\frac{3}{2}\frac{(1+\sigma_b)}{NPSE}\right\}^{\frac{1}{2}} \tag{8.15}$$

and:

$$\text{Speed, } N = 35.2\sqrt{\frac{1-\dfrac{D_h}{D_t}}{1+\dfrac{D_h}{D_t}}}\;\frac{NPSE^{\frac{3}{4}}}{\sqrt{Q}}\left(\frac{1}{(1+\sigma_b)\sigma_b{}^2}\right)^{\frac{1}{4}}$$

where:

$$\sigma_b = \frac{(P_1 - P_v)}{\dfrac{1}{2}\rho W_1^2} = \text{blade cavitation coefficient}$$

Q is the flow rate in m³/s, N is the speed in rpm, ω is the angular velocity in rad/s, and NPSE (Net Positive Suction Energy) = g.NPSH, where g is the acceleration due to gravity in m/s².

The ratio of hub diameter to tip diameter at the inlet, i.e., D_h/D_t, is to be kept as low as possible and Pearsall recommends a value of 0.3.

2. The approach to pump suction is streamlined so that liquid approaches the impeller with a gradually increasing velocity.
3. Small blade inlet angle, β_1, is preferred to keep the NPSH requirement low. This is achieved by a large impeller eye diameter. A practical lower limit on β_1 is ~15°. Lowering β_1 further reduces the pump efficiency.
4. Providing a small degree of pre-whirl at the impeller inlet in the same direction as the direction of rotation of the impeller. Pre-whirl is produced by designing the impeller for larger than the nominal discharge. Pre-whirl is also achieved by providing guide vanes at the impeller inlet.
5. Using fewer blades for low specific speed pumps, and more blades for high specific speed pumps.
6. Using blades of double curvature, i.e., blades with inlet edge extended into the inlet (mixed flow impeller).
7. Shortening the inlet edge of alternate blades in impellers with even numbers of blades.
8. Sharpening and reducing the blade thickness at the inlet.

9. Increasing the inlet width of the impeller.
10. Using a double suction impeller.
11. Providing inducer upstream of the impeller thus imparting energy to the liquid and preventing vapour bubble inception before energy additions begins in the impeller.
12. Improving the surface finish of the impeller thereby delaying the inception of cavitation bubbles.

8.7 CAVITATION CRITERIA FOR CRITICAL APPLICATIONS

Cavitation in a pump manifests in the form of acoustic noise, visible bubbles, erosion damage, and head drop depending on the extent of development of the process. This is illustrated in Figure 5.4 in Chapter 5 and Figure 8.1 in the present chapter. Conventionally the 3% head drop, at a specific flow rate, is considered as an indicator of classical cavitation and the NPSH at this flow rate is considered as the NPSHR of the pump. This value of NPSH, however, corresponds to a developed degree of cavitation, as seen in Figures 5.4 and 8.1, and damage due to cavitation cannot be ruled out. The simplistic solution to the problem of cavitation damage is to provide an adequate margin on the $NPSHR_{3\%}$. As explained in Section 8.1, a margin of 4 times the $NPSHR_{3\%}$ may be the least required to ensure no damage is produced especially in the case of high energy pumps and this can result in bulky and expensive equipment. For critical applications such as the main coolant pumps of liquid metal reactors several alternative cavitation criteria have been explored based on acoustic noise, visible bubbles and $NPSHR_{0\%} + x$ (a margin) [4]. The selected criterion is validated through studies on scaled down models in water. Ref. [21] discusses in detail the tests done in water on a 1/2.75 scaled model of the PFBR primary sodium pump and interested readers can consult the same for more details. These tests are useful in: (i) identifying cavitation susceptible areas in the impeller; (ii) estimating the cavity length in the prototype pump under plant NPSH conditions using the measured cavity size from the model tests; and (iii) estimating the life of the prototype pump impeller using empirical correlations [22].

8.8 SUMMARY

Cavitation performance of centrifugal sodium pumps for reactor applications is important because (i) the pump is to be compact to reduce capital cost (ii) the cavitation damage produced in liquid sodium is intense and is therefore to be avoided, and (iii) removal and replacement of pump hydraulic parts is a time consuming and costly affair affecting reactor availability. This chapter briefly discusses the phenomenon of cavitation, the thermodynamic effect on cavitation and liquid properties that influence it, the types of cavitation, and their damage locations and the important design parameters that affect pump cavitation performance.

NOTE

1 Jakob number is the ratio of the sensible heat to the latent heat absorbed during the liquid to vapour phase change process.

REFERENCES

1. M.S. Plesset, A.T. Ellis, On the mechanism of cavitation damage, *Transactions of ASME* 77, 1055–1084, 1955.
2. H. Addison, *Centrifugal and other Rotodynamic Pumps*, Chapman and Hall, 1966.
3. F.G. Hammitt, Detailed cavitation flow regimes for centrifugal pumps and Head vs NPSH curves, Report no. UMICH 01357-32-I, University of Michigan, December 1974.
4. C.E. Brennen, *Hydrodynamics of Pumps*, Concepts ETI Incorp. & Oxford University Press, 1994.
5. Proceedings of the IWGFR Specialists' Meeting on Cavitation Criteria for Designing Mechanisms Working in Sodium Application to Pumps, Interatom GmbH, Federal Republic of Germany, 28–29 October 1985.
6. C. Gongwer, A theory of cavitation flow in centrifugal pump impellers, *Transactions of ASME*, 63, 29–40, n.d.
7. F.M. Mitenkov, E.G. Novinsky and V.M. Budov, Main Circulation Pumps for Atomic Power Stations, edited by F.M. Mitenkov, member, Academy of Sciences, USSR, 2nd Edition, ENERGOATOMIZDAT, Moscow, 1990.
8. R.T. Knapp, J.W. Daily and F.G. Hammitt, *Cavitation*, McGraw Hill, 1970.
9. B.K. Sreedhar, S.K. Albert and A.B. Pandit, Cavitation Damage–A Review, *Wear* 372–373, 177–196, 2017.
10. H.A. Stahl and A.J. Stepanoff, Thermodynamic Aspects of Cavitation in Centrifugal Pumps, *ASME Journal of Basic Engineering*, 78, 1691–1693, 1956.
11. A.J. Stepanoff, Cavitation in Centrifugal Pumps with Liquids Other than Water, *ASME Journal of Engineering for Power*, 83(1), 79–89, 1961.
12. A.J. Stepanoff, Cavitation Properties of Liquids, *ASME Journal of Engineering for Power*, 86, 195–199, 1964.
13. L.W. Florschuetz and B.T. Chao, On the Mechanics of Vapour Bubble Collapse, *Journal of Heat Transfer, Transactions of ASME*, 87(2), 209–220, May 1965.
14. R. Garcia, Comprehensive Cavitation Damage Data for Water and Various Liquid Metals Including Correlation with Material and Fluid Properties, PhD Thesis, University of Michigan, 1966.
15. R. Garcia and F.G. Hammitt, Cavitation Damage and Correlations with Material and Fluid Properties, *Journal of Basic Engineering*, 89(4), 753–763, December 1967.

16. H.S. Preiser, A. Thiruvengadam and C.E. Couchman III, Cavitation Damage in Liquid Sodium, Technical report 285-1, NASA CR-54071, April 1964.
17. A.J. Stepanoff and H.A. Stahl, Cavitation Criterion for Dissimilar Centrifugal Pumps, *ASME Journal of Engineering for Power*, 84(4), 329–336, October 1962.
18. S. Lazarkiewicz and A.T. Troskolanski, *Impeller Pumps*, Pergamon Press, 1965.
19. R.K. Turton, *Rotodynamic Pump Design*, Cambridge University Press, New York, 1994.
20. I.S. Pearsall, Cavitation, M&B Monograph ME/10, 1972.
21. S.G. Joshi, A.S. Pujari, R.D. Kale and B.K. Sreedhar, Cavitation Studies on a Model of Primary Sodium Pump, *Proceedings of FEDSM02, The 2002 Joint US ASME European Fluids Engineering Summer Conference*, July 14–18, 2002, Montreal, Canada.
22. J.F. Gulich, *Centrifugal Pumps*, Springer, 2008.

Chapter 9

Main coolant pumps for water-cooled reactors

9.1 INTRODUCTION

In the preceding chapters, we have covered all aspects of design, construction, and operational issues of sodium coolant pumps used in fast nuclear reactor systems. This chapter briefly discusses the features of main coolant pumps (MCP) of water-cooled reactors, emphasising MCPs for Indian Pressurised Heavy Water Reactors (PHWRs). A brief examination of coolant pumps of light water reactors such as PWRs (VVERs) is also included.

The discussion, however, is restricted only to pumps with seals in contrast to hermetically sealed pumps.

The purpose of including a discussion on water-cooled reactor pumps is to bring together essential aspects or features of such pumps for the benefit of readers who are conversant with water-cooled reactors (including heavy water reactors) but need more access to information on such pumps. The authors think this will be of interest to engineers working with nuclear power plants in India.

The Indian PHWR series of heavy water moderated and cooled reactors are of capacity 220 MWe, 540 MWe, and recently of 700 MWe. The early reactors from the 1970s to the early 21st century were mainly of 220 MWe capacity, and only two plants of 540 MWe were constructed and commissioned in early 2000 at Tarapur (TAPS 3 and 4). Currently, several plants of 700 MWe capacity, which are basically upgraded versions of 540 MWe designs, are under advanced stage of construction and in June 2023 the Kakrapar Atomic Power Plant (KAPP-3) went commercial.

The MCPs for the early 220 MWe plants were largely imported, and those from Narora nuclear power plant onwards were supplied by KSB (India) and supported by KSB Germany. The main coolant pumps of 540 MWe and 700 MWe plants are of the same design/capacity and are from KSB, India.

9.2 MAIN COOLANT PUMP FOR PHWR

In a 220 MWe plant, each reactor outlet header is connected through two parallel 400 mm NB outlets to two steam generators (SG). The outlet of

DOI: 10.1201/9781003460350-9

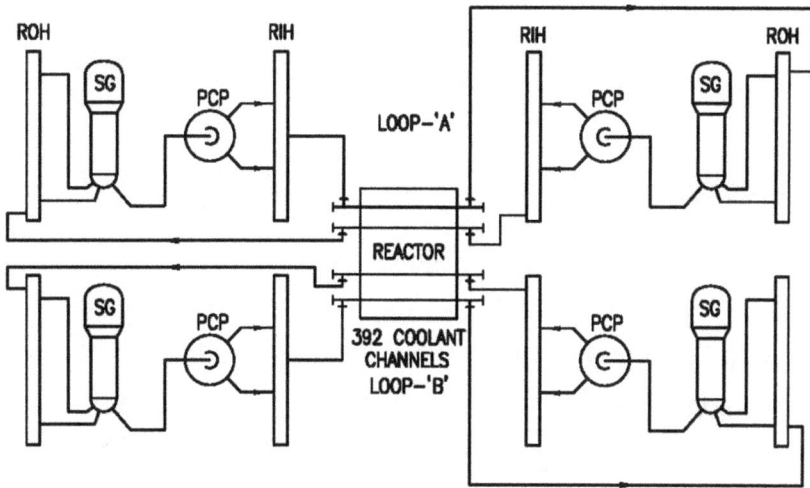

Figure 9.1 Main coolant circuit in PHWR. PCP – Primary Coolant Pump; RIH – Reactor inlet header; SG – Steam generator; ROH – Reactor Outlet header.

each SG is connected by 400 NB pipes to the corresponding pump suction. The two primary circulating pumps are mounted in parallel with their discharge nozzles connected to corresponding reactor inlet headers. Figure 9.1 illustrates this arrangement. The coolant flow rate in the main circuit is arranged in a figure of eight configuration. From the design of the Kaiga NPP onwards, no valves are provided in the main coolant circuit.

9.2.1 MCP for 220 MWe PHWR [1]

Each pump is designed to provide the rated flow of 3560 m³/h of heavy water at head of 178 m at 245°C. The pump (Figure 9.2) is a vertical single-stage unit with a radial impeller, volute casing, axial bottom inlet (suction), and horizontal radial discharge. The operating speed of the pump is 1481 rpm, and it is driven by an electric motor with a power rating of 2,800 kW. A single-piece flywheel mounted on the non-drive end of the motor shaft provides the inertia for safe coast down in the event of a pump trip.

Three hydrodynamic seals operating in series prevent leakage of high-pressure radioactive heavy water from the pump.

The seal injection system comprises a jet pump, a high-pressure cooler, and a cyclone separator. The seal system is capable of operating both with and without external injection. Low differential pressure across the seal faces improves the reliability of the seals; the seals are, however, capable of operating under full system pressure.

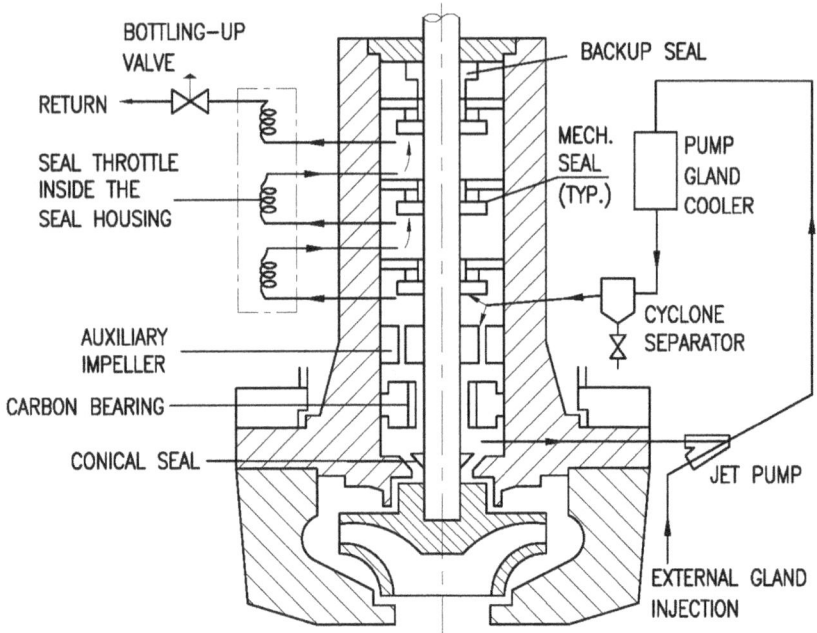

Figure 9.2 Sketch of shaft seal system of MCP of 220 MWe PHWR.

The seal cavity water is maintained cool and clean under normal/transient conditions and pump-stopped conditions with or without external injection. In the event of failure of the third seal, leakage is limited during pump coast down and pump-stopped conditions, using a backup spring-loaded vapour seal. A metal-to-metal conical seal between the pump shaft and carbon-bearing housing makes it possible to carry out seal maintenance without draining of pump casing. The shaft seal system is shown in Figure 9.2.

Table 9.1 gives salient design parameters of the main coolant pump of the 220 MWe plant.

The rotating assembly is a three-shaft system. The pump shaft is supported at the bottom by a hydrodynamic carbon-bearing located above the impeller. The thrust load is supported by dedicated oil-lubricated, tilted pad-type thrust bearing mounted on an intermediate shaft. A spacer coupling connects the pump shaft and the intermediate shaft. A flexible coupling connects the intermediate shaft to the motor shaft. The motor shaft is supported by oil-lubricated, tilted pad-type radial and thrust bearings.

The arrangement facilitates the inspection/replacement of bearings and seals without motor removal (in contrast to sodium pumps, where the motor is to be removed to replace the seals and bearings).

The pumps are supported from the top by two variable support spring hangers at the motor stool/top flange location and sliding support at the pump suction elbow location. About 80% of the weight of the pump-motor

Table 9.1 Design parameters of primary coolant pump of 220 MWe plant

Type	Vertical centrifugal, single stage
Rated flow	3560 m³/h
Rated head	178 m (design) 184 m (actual)
Inlet pressure	83.4 bar (g) (85 kg/cm² (g))
Design pressure	110 bar (g) (112.5 kg/cm²(g))
Fluid temperature	249°C
NPSH available at 249°C	33.1 bar(g) (45 kg/cm²(g))
NPSH required	3 m (cold)
Speed	1490 rpm
Inlet	400 mm NB ×31 mm thick 90° LR Elbow, BW, CS
Outlet	400 mm NB ×31 mm thick 90° LR Elbow, BW, CS
Specified minimum flow rate after loss of power supply	70% at 5 secs. and 6% at 50 secs.
Run-down method	By use of flywheel
Motor type	Squirrel cage induction
Motor rating	2,800 kW, 6.6 kV, 50 Hz

assembly is supported by the spring hangers and the remaining by the bottom support, thus minimising vibrations during operation. This arrangement permits the sliding of the pump during the heating and cooling of the primary heat transport system, thus minimising the thermal stresses.

9.2.2 MCP for 500 MWe PHWR [2]

The MCP is a close coupled, vertical, single-stage, radial impeller with single bottom entry and double discharge (downward) from a symmetrical volute casing. Some of the important design improvements in these pumps, when compared to the pumps in the 220 MWe plants (described earlier), are as follows:

(i) Materials having intrinsically high-notch toughness (impact strength) are used for pressure boundary.
(ii) Pumps are designed for a life span of 40 years, operating at the rated condition for 8,000 h annually with a total of 4,000 starts and stops during design life.
(iii) Each pump is designed to operate at ~66% rated flow when one pump in the loop stops, thus allowing part load operation of the reactor.
(iv) A brake is provided in the motor to prevent turbining of stopped pump/pumps (i) due to flow from other pump/pumps in the same loop (ii) during thermo-syphoning of coolant flow on total loss of pumping power.

(v) Station transients and associated stress analysis covering service limits (A, B, C, & D) are identified in accordance with ASME Sec. III.

(vi) In addition to the three mechanical seals in series and the vapour seal, an additional stationary backup seal is provided before the vapour seal to prevent uncontrolled leakage of heavy water to the reactor building atmosphere in the event of failure of the third mechanical seal. The seals are capable of operation without external gland cooling for an extended period. The backup seal is designed to withstand the full system pressure and increases the life of the third mechanical seal by reducing the pressure differential across the seal when the gland system is boxed up during hot standby condition or hydrostatic testing. The backup seal is usually dormant during pump operation (locked by spring/piston) and is actuated by nitrogen gas pressure on failure of the third mechanical seal.

(vii) The coupling is sized to facilitate quick removal of the seal assembly or the motor. The coupling is also designed to have sufficient axial/radial stiffness to transmit axial thrust to the upper thrust bearing while retaining enough flexibility to accommodate the misalignment of pump/motor shafts.

(viii) The coupled shafts of the pump and motor rotate in a three-journal bearing system. The bottommost bearing is heavy water lubricated carbon (radial) bearing (on the pump shaft), while the top radial bearing and thrust bearing are both on the motor shaft. The thrust and radial bearings on the motor shaft are oil-lubricated tilted pad-type bearings, which are suited for reverse direction also. With this bearing arrangement, it is possible to eliminate the thrust bearing on the pump shaft thus achieving a two-shaft design instead of the three-shaft design of the pumps in 220 MWe plants.

Self-lubrication of the thrust and guide bearings is achieved during normal operation. The rotating thrust collar forces the oil flow through bore holes in the thrust collar head to the bearing pads and the oil cooler. There is no need for an external oil supply as the oil sump and circulation are integral to the bearing. The arrangement bears similarity with the top thrust/journal bearing of the SNR-300 sodium pump, which is also a self-lubricated system. While this is true during normal pump operation, during startup, coasting at low speeds, and shutdown, however, the upper thrust-bearing shoes must be separated from the thrust collar against a static upward thrust. The bearing jacking oil system ensures this and establishes an oil film on the thrust pads. The system consists of motor-driven jacking oil pumps, necessary valves, filters, piping manifolds, and pressure/flow switches.

(ix) Additional requirements related to the examination of pressure boundary welds, over and above that mandated by code, are included. The Heat Affected Zone (HAZ) and the adjacent base material will be examined by Magnetic Particle (MP) and Ultrasonic Testing (UT) as part of pre-service inspection (PSI) after final heat treatment and hydrostatic test.

(x)   The flywheel, mounted between the two bearings of the motor, complies with the requirements of US NRC Regulatory Guide 1.14 (Rev.1, Edition August 1975) concerning design, material, fabrication, testing, and inspection. The flywheel material (SA 508 C14) is of closely controlled quality (impurity level 1.0 according to ASTM E-45) and is tested for fracture toughness and tensile properties to confirm the values assumed for analysis. Flywheel analysis includes predicting the critical speed for non-ductile fracture and fatigue crack growth behaviour of radial cracks at the inner surface during the intended service life of 40 years coupled with 4,000 starts/stops. The objective is to ensure that indications accepted during ultrasonic testing will stay within the critical length during the service life.

Figure 9.3 is a sketch of the main primary coolant pump in the 500 MWe plant.

Table 9.2 gives the salient design parameters of the main coolant pump of the 500 MWe plant.

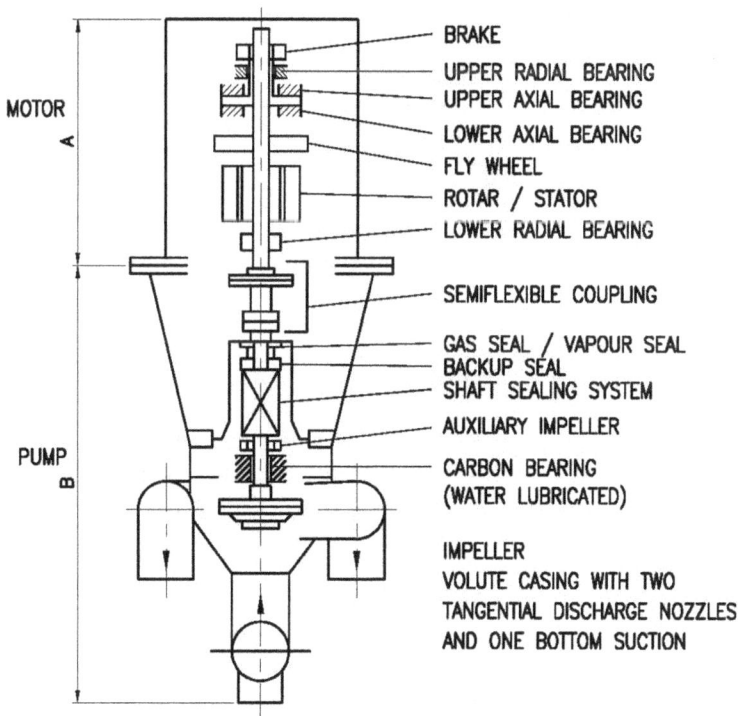

Figure 9.3 Schematic arrangement of main primary coolant pump in the 500 MWe Plant.

Table 9.2 Design parameters for primary coolant pump of 500 MWe plant

Design classification	ASME Section III Class I
Seismic classification	Seismic Cat. I should remain functional after OBE and no breach of pressure boundary as a result of SSE.
Design pressure	123.6 bar (126 kg/cm²)
Design temperature:	310°C
Operating pressure	99.0 bar (101 kg/cm²)
Operating temperature	260°C
Speed	1,487 rpm
Operating head	215.4 m
Rated flow	8,400 m³/h
Maximum flow rate	11,760 m³/h
Hydrostatic test pressure	154.5 bar (157.5 kg/cm²)
Hydrostatic test temperature	20°C
Electrical load	6,000 kW, 6.6 kV, 3 phase
	Material
Volute casing	SA 352 Gr. LCB
Suction elbow	SA 420 Gr. WPL 6
Discharge elbow	SA 420 Gr. WPL 6
Shaft	X5 CR NI 134 (SA 182 Gr. F6 NM)
Impeller	GX5 CR NC 134
Seal housing	SA 182 Gr. F6 NM
Seal cover	SA 182 Gr. F304
Seal faces:	
Stationary	Silicon Carbide
Rotating	Silicon Carbide
Motor Stand	SA 516 Gr. 70

9.3 MAIN COOLANT PUMP FOR VVER REACTOR [3]

The largest MCP designed and built in Russia is the primary coolant pump for VVER-1000 MWe nuclear power plant. It has a capacity of 20,000 m³/h and is driven by a 4800-kW induction motor. The VVER is a pressurised water-water reactor, and the Kudankulam Nuclear Power Plant (KNPP) at Kudankulam is based on the VVER-1000 MWe design.

The MCP in the 1000 MWe reactor (in Russia) has a wholly cast volute casing, a removable part, and an electric motor with a top baffle fin, a bearing oil system, and a cooling water system. The welded-cast annular frame with support lugs is the load bearing structure of the pump. The pump is installed with its lugs bolted to supports mounted on the foundation. The supports are provided with bearings that permit thermal displacement of the pump unit under piping forces. The volute casing enclosing the pump hydraulics is situated under the biological shielding while the pump top bearing, seals, and other removable parts, as well as the motor, are placed inside a box structure above the foundation to be accessible for inspection during pump operation or maintenance.

The pump rotor assembly is supported by a two-bearing system. The bottom hydrostatic bearing is supplied with high-pressure water from an auxiliary impeller; the top radial cum axial (thrust) bearing is oil-lubricated. The pump shaft and motor shaft are connected using splined half couplings and a torsion shaft. A flywheel is mounted at the lower end of the motor shaft. The anti-reverse rotation device mounted at the top end of the pump shaft, above the thrust bearing, prevents reverse rotation of the idle pump in the event of a bypass of flow (from the operating pump) through the non-return valve of the stopped pump.

Torque from the motor is transmitted to the pump rotor through a torsion coupling.

Radiation streaming is prevented by installing a carbon steel ring of 315 mm thickness above the volute casing. Streaming through the annular space between this ring and the volute casing is prevented by a covering ring (thickness 300 mm). The covering ring also provides a firm base for fixing the pump support lugs.

The sealing block consists of three stages of hydrostatic end face seals. Pure water to the seal is supplied by dedicated high-pressure (HP) feed pumps through filters and hydro cyclones that ensure the removal of particles greater than 10 μm size. Under emergency conditions water from the impeller outlet is supplied through a cooler to the seal. The sealing pairs are made from silicon-graphite, while the remaining parts of the block are from austenitic steel.

The pump has an extended coasting down time, with the flow rate reducing by only 2.7 times 30 secs after power supply to the motor is cut off. The pump is hydraulically tested along with the primary circuit (piping) at 25MPa pressure with coolant water temperature between 50°C and 130°C.

9.4 MAIN COOLANT PUMP FOR NUCLEAR POWER PLANT (NPP) AT LOVIISA, FINLAND [3]

This pump is developed by a Finnish firm for use in their NPP, which is based on VVER-440 Russian nuclear plant. However, this MCP has some specific features which make its design different from similar MCPs. A major difference is in the hydraulic layout, which in this pump uses a side entry, top suction, and bottom discharge impeller. The arrangement does not need high-pressure sealing as in other pumps.

The pump is driven by a squirrel cage, induction motor of 1,300 kW capacity, and consists of a housing and a removable part (with mixed flow impeller and diffuser). The housing has a side intake and bottom discharge pipes. This arrangement simplifies the joining of pipelines to the housing and helps to reduce the mass and overall size of the housing and the size of the main joint. The sealing is also simplified as it is connected to the inlet region of the pump.

The pump shaft is connected to the motor shaft using a rigid coupling resulting in a combined rotor assembly that rotates in a three-bearing system. The critical speed of the rotor is 1.25–1.3 times higher than the operating speed. A hydrodynamic sleeve bearing is used as the bottom bearing of the shaft. The bearing is lubricated and cooled by water circulated through an independent circuit using a separate auxiliary impeller mounted on the same shaft. The motor has two oil-lubricated, frictionless bearings, one of which also takes axial load transferred by the pump shaft through the rigid coupling (with annular keys). In the coupling itself, there is a 370 mm gap between the ends of the two shafts, which allows the replacement of the seal and bearing of the pump without disturbing the electric motor. The pump cover is bolted to the housing using 24 stud bolts. Leak tightness is achieved using two spirally wound gaskets with graphite filling. The shaft seal has three stages: two hydrostatic end faces which distribute equally the total pressure drop and a top end face seal designed for 0.5MPa.

The material of construction of components in contact with the hot coolant is austenitic stainless steel. Assemblies which require special attention during manufacture include the bottom cast housing which is spherical in shape.

The electric drive motor is fitted with a flywheel to ensure gradual coast down of the pump when motor is tripped. This allows reliable cooling of the reactor in all operating modes. Under the flywheel a combination of ring-shaped electromagnet and force measuring strain gauge is provided to measure the axial force acting on radial cum axial bearing. It also permits unloading of the bearing by adjusting the electromagnetic force based on the force measured by the strain gauge.

The pump assembly is fixed tightly (rigidly) on the frictionless support by means of hinged rings. Three rings with four hinges are fixed to the pump housing to form a hinged suspension. Three supports installed on brackets are fixed to the outer ring of the suspension. Each support has damping springs and rollers on which the pump slides. The top ring restricts movement of MCP in the case of rupture of pipeline (LOCA) of the multiple forced circulation circuit. These pumps have been successfully operating in I and II blocks of the NPPs at Loviisa, Finland.

9.5 SUMMARY

This chapter gives an overview of the construction of main coolant pumps in water-cooled reactors. A major difference of these pumps vis-à-vis sodium centrifugal pumps is that these pumps operate at high pressures which are essential to keep the coolant (light water/heavy water) in the liquid state at the operating temperature. As a result the available NPSH is significant even at the high operating temperature and the pumps are of short length (short shaft construction). Other important differences in the operating conditions of light/heavy water pumps when compared to that of sodium pumps include relatively lower operating temperature and capacity regulation by valve throttling instead of speed regulation as in sodium pumps.

REFERENCES

1. S.S. Sharma, S.G. Mhatre and M.M. Manna, Upgradation of design features of primary coolant pumps of Indian 220 MWe PHWR, Pumping Equipment in Nuclear Industry and Thermal Power Plants, Proceedings *of National Symposium*, BARC, Bombay, February 24–25, 1994.
2. V. Misri, C.N. Bapat and V.K. Sharma, Primary coolant pumps for 500 MWe Pressurized Heavy Water Reactor (PHWR) – Specification, Design, Manufacturing and Testing Requirements – An Overview, Pumping Equipment in Nuclear Industry and Thermal Power Plants, Proceedings *of National Symposium*, BARC, Bombay, February 24–25, 1994.
3. F.M. Mitenkov, E.G. Novinsky and V.M. Budov, Main Circulation Pumps for Atomic Power Stations, edited by F.M. Mitenkov, member, *Academy of Sciences, USSR*, 2nd Edition, ENERGOATOMIZDAT, Moscow, 1990.

Chapter 10

Sodium pumps of the future

10.1 INTRODUCTION

The previous chapters described the features of main sodium centrifugal pumps in experimental, prototype, and commercial reactors, the considerations in designing these pumps, their fabrication requirements, the features of test facilities, and the operating experience of primary and secondary sodium centrifugal pumps. We also dwelt on the phenomenon of cavitation and its importance in the design and operation of sodium pumps and described, albeit briefly, the main coolant pumps in water-cooled reactors.

In this chapter, we foresee features of future centrifugal coolant pumps in sodium-cooled reactors.

The important consideration in designing sodium-cooled reactor equipment for future reactors will be economy, robust performance, minimum downtime/maintenance, and long life. For sodium centrifugal pumps, the challenge of meeting these requirements entails improvements in the hydraulic and mechanical design.

10.2 DESIGN FEATURES OF FUTURE CENTRIFUGAL PUMPS

10.2.1 Pump hydraulics

Sodium-cooled reactors are low-pressure systems; therefore, the available NPSH, especially in the primary radioactive circuit, is modest. This limited available NPSH restricts the pump design speed, thereby restraining efforts to minimise the pump impeller diameter or lateral dimension and the capital cost of both the pump and the reactor (especially so for the pool-type design as the primary coolant pump is inside the main vessel and therefore has a direct bearing on the vessel diameter).

Increasing the efficiency of the hydraulic design in both the inlet flow path and the impeller suction is crucial for effectively delaying the inception and growth of cavitation bubbles. By exploring design options that promote the break-up and dispersion of vapor bubbles, the clustering and uncontrolled expansion of the bubbles can be avoided. Additionally, it is essential to

DOI: 10.1201/9781003460350-10

prevent bubble collapse near the bounding surface to avoid the deleterious effects of cavitation damage. Thankfully, the continual advancement of computational modelling tools and techniques promises to realize these challenging yet attainable solutions.

Surface treatment/hardfacing of cavitation-susceptible regions of the pump (e.g., impeller suction face) can improve resistance to cavitation damage [1–3] facilitating operation with moderate degrees of cavitation. Ni and Co-based hardfacing alloys are likely candidate materials for this purpose (although Co-based alloys have higher resistance to cavitation damage [4], the presence of Co-60 in activated components can pose a hazard during maintenance and repair, thus limiting its use; Ni-based hardfacing alloys do not have this disadvantage and are therefore preferred in radioactive applications). The challenge in implementation, however, is the closed configuration of the impeller (blades enclosed by front and back shrouds). In this context, the possibility of a fabricated/welded construction that offers better surface finish can be explored.

An alternative option is to use hardfaced inducers mounted upstream of the impeller on the pump shaft. The inducer's open design is more conducive to hardfacing, making it more resistant to cavitation damage. The inducer adds energy to the liquid and prevents cavitation at the impeller inlet, permitting higher operating speed, and making the pump more compact.

10.2.2 Pump shaft

The pump shaft in all sodium pumps is purposefully designed with a composite geometry, featuring both hollow and solid sections. This particular design serves two crucial purposes. Firstly, it ensures that the impeller is well-submerged, meeting the NPSH requirements, which is vital for the pump's optimal functioning. Secondly, the composite geometry helps maintain a sufficient margin between the operating speed and the first critical speed of the shaft, without adding excessive weight to it. However, this geometry results in increased fabrication cost. An alternative option for the future is the use of a supercritical shaft design in which the pump is designed to operate above the critical speed. This approach allows for the use of a less bulky shaft. However, difficulties associated with part load operation where the running speed may be close to the critical speed are to be addressed.

10.2.3 Top bearing and seals

10.2.3.1 Arrangement to replace the bearing/seal unit without removing the motor

In the present arrangement of the pump and the motor, replacing the top bearings and seals requires the disconnection of the flexible coupling and removal of the motor assembly for access to the bearings and seals.

This arrangement can be modified to include additional axial space between the motor and pump shafts' ends with an intermediate spacer shaft and flexible coupling to interconnect the pump and motor shafts. In the event of any maintenance requirement, the flexible coupling and intermediate shaft may be disconnected, the motor moved away from the pump shaft centreline, and the bearings and seals inspected/replaced. This can result in a significant reduction in maintenance time. However, the complexity of the dynamics of the modified system is to be addressed beforehand through experimental studies.

10.2.3.2 Oil-less bearings and seals

Conventional sodium centrifugal pumps use oil lubricated radial and thrust bearings to support the rotating assembly above the argon cover gas space, along with oil-cooled mechanical seals to contain the cover gas above the free surface of sodium. Despite the provision of design features (e.g., catchpot of adequate capacity to contain the leaking oil) and operational procedures (e.g., oil level measurement in catchpot, periodic draining, etc.) there have been instances of contamination of sodium due to oil leak from the bearing/seal system (e.g., the oil leak into the primary circuit of the Prototype Fast Reactor (PFR), in UK).

Oil-less bearings such as foil air/gas bearings and active magnetic bearings (AMB) have made significant progress in the last four decades. AMB's are preferred over foil bearings for high load and lower speed requirements [5]. AMB's are in use in large turbomachinery like gas compressors, turboexpanders, multistage boiler feed pumps, etc. [6]. They have been designed for the helium turbine generator system [7] and turbine compressor rotor [8] of the 10 MW high-temperature gas-cooled reactor (HTR-10GT) in China. Some of the distinct advantages of AMB, other than oil-free use, are: (i) absence of wear; (ii) reduced friction; (iii) possible unbalance compensation; (iv) adjustment of bearing stiffness and damping by tuning the system; (v) use as a force measurement tool to measure hydraulic forces on pump rotor and integration of fault monitoring; and (vi) use as a fault diagnostic system with the turbomachine [9]. The most attractive feature of AMB is its ability to measure and internally process signals to optimise the state of the machine (Smart machine) [10].

Oil is used in the mechanical seals also to lubricate and cool the seal faces. In place of these conventional seals, the ferrofluid seal provides an oil free option. A ferrofluid is a stable suspension of nanosized magnetic particles coated with a stabilising agent to prevent agglomeration in a carrier liquid. In the presence of a magnetic field gradient, the fluid responds as a homogenous magnetic substance and moves to the region of the high magnetic field.

This response to the external magnetic field is exploited to produce the sealing effect. In its simplest form, the ferrofluid seal consists of a permanent magnet flanked on either side by magnetically permeable pole pieces. The flux generated by the magnet passes through the pole pieces and the magnetically permeable shaft (or shaft sleeve) to complete the magnetic circuit. The resulting concentrated magnetic flux in the radial gap between the pole pieces and the shaft causes the ferrofluid injected into this gap to function as a 'liquid O-ring'.

Dynamic ferrofluid seals have been used in high vacuum applications and for sealing with pressure differential greater than 1.7 MPa; only mechanical considerations limit the capability of the seal to hold pressure. Dynamic ferrofluid seals have been used up to surface speeds of 10 m/s. Radiation-resistant ferrofluid seal has been used in a CAT scanner at rotational speeds of 3000 rpm–6000 rpm to seal vacuum (of the order of 10^{-6} bar–10^{-7} bar) [11].

An oil-free centrifugal sodium pump can be realised by replacing the existing oil-cooled thrust/radial bearings and mechanical seals with a combination of AMB and ferrofluid seals.

10.2.4 Integrated condition monitoring

This section explores the possibilities of minimising instrumentation hardware in future sodium pumps by using the pump components themselves for condition monitoring. Some such possibilities are given below.

10.2.4.1 Conditioning monitoring using AMB

In conventional centrifugal sodium pumps, the health of the pump is monitored through independent measurement of various pump parameters such as flow rate, speed, and vibrations monitoring. Active Magnetic Bearings (AMB) provide the pumps of the future with not just an oil free bearing solution but an integrated condition monitoring system. Some potential applications include:

(i)   Vibration analysis: One of the primary benefits of using AMBs is that the sensors in the AMB system can be used to measure and record the vibrations of the rotating shaft.
(ii) Rotor position and displacement monitoring: Active Magnetic Bearings can precisely control the position and displacement of the rotating shaft. This capability allows for real-time monitoring of the shaft's behaviour during operation thereby enabling early identification of problems such as rotor rub, rotor eccentricity or rotor instabilities.
(iii) Current and voltage monitoring: Monitoring the current and voltage supplied to the AMBs can provide information on the forces acting on the rotor. Any abnormal fluctuations or irregularities in the supplied power can indicate potential issues with the machinery's operation or the magnetic bearing system itself.
(iv) Temperature monitoring: Temperature sensors can be integrated into the AMB system to monitor the operating temperature of the pump shaft and flag unusual temperature variations.

(v) Data analytics: Using advanced data analysis and machine learning techniques the historical data collected from the AMB sensors can be analysed for patterns and used to anticipate and prevent failure.

(vi) Remote monitoring and diagnostics: The real-time condition monitoring data from AMBs can be transmitted and monitored remotely by off-site experts facilitating timely and effective analysis of the health of the pump.

10.2.4.2 Cavitation monitoring using eddy current probe thimble

The conventional method of cavitation monitoring involves measuring the 3% pump head drop at the operating speed and flow rate. However, in reactor centrifugal pumps, this approach is avoided because (i) the pressure tappings for pump head estimation are potential sources of leak, and (ii) 3% head drop represents advanced stage of cavitation and operating the pump close to this NPSH value is not desirable. For future pumps, various alternative measurement techniques can be employed. For example, in the PFBR primary sodium pump, an eddy current probe located at the impeller outlet is used to measure the flow rate. The sensor is enclosed in a pocket to protect the sensor from sodium and facilitate its in-situ replacement. In future sodium pumps, this pocket can also function as a waveguide, facilitating the monitoring of high-frequency cavitation noise signals. Detection and analysis of these signals will facilitate cavitation detection at a much earlier stage than that possible with 3% head drop criterion.

10.3 INTEGRATED PUMP AND INTERMEDIATE HEAT EXCHANGER

In the early years of the new millennium, the Japan Nuclear Cycle Development Institute carried out a feasibility study to identify economically competitive fast reactor systems for the future. Different options were studied based on criteria such as safety, effective utilisation of resources, reduction of environmental burden, enhancement of nuclear non-proliferation, and finally, economic competitiveness. Cost reduction was proposed through the design of compact reactor structures, shortening of piping length, reduction in the number of loops, and integration of components. This was to be realised through a change of material from austenitic stainless steel to 12Cr steel (higher strength and lower coefficient of thermal expansion) and adoption of innovative technologies such as advanced elevated temperature structural design standards, three-dimensional seismic isolation and re-criticality free technology [12].

Relevant to our discussion is the integration of components. The primary pump and the intermediate heat exchanger (IHX) could be integrated with the pump located centrally inside the IHX. The integration of the pump and IHX, shown in Figure 10.1, yields cost benefits through the elimination of: (i) the pump tank and its guard vessel (in a loop-type reactor); (ii) the

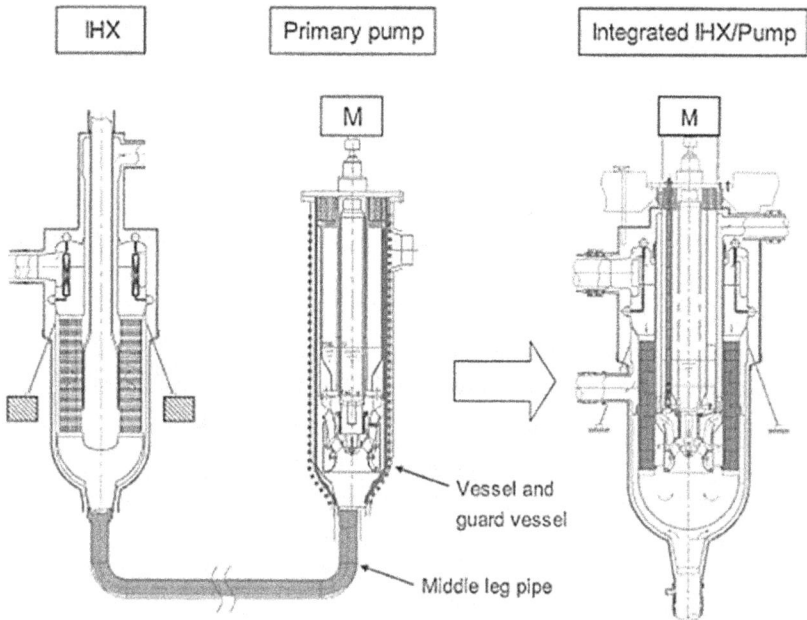

Figure 10.1 Integrated primary pump and IHX [13]. (Copyright © Atomic Energy Society of Japan, reprinted by permission of Taylor & Francis Ltd., http://www.tandfonline.com on behalf of Atomic Society of Japan).

connecting piping between IHX and pump; and (iii) reduction of building size.

Although this could result in a larger tube sheet diameter of the IHX, the high strength and low thermal expansion coefficient of 12Cr steel are expected to ensure structural integrity (Figure 10.1) [12].

The integration, however, may require additional studies to understand technical issues such as: (a) the effect of transmission of pump vibration to heat exchanger tubes and its effect on the wear of heat exchanger tubes; (b) the effect of gas entrainment on flow rate; (c) effect of cavitation on flow rate; (d) design to minimise heat load to pump top bearing; and (e) effect of gas convection on thermal deformation of shaft and casing. In this context, tests have already been conducted on a ¼-scale model of the integrated pump and IHX to study the vibration of the assembly and prevention of gas entrainment due to the rotating shaft [13]. The integrated pump and IHX concept is an attractive proposition for loop-type reactors where the pipe length between the pump and IHX is substantial.

10.4 SUMMARY

The mandate for cost-competitive power generation will require the main sodium centrifugal pumps of the future to be more compact and capable of operating for long periods with minimal maintenance. This chapter attempts to peer into the future and reconstruct the important details of such a pump.

REFERENCES

1. J.H. Boy, A. Kumar, P. March, P. Willis and H. Herman, Cavitation and Erosion Resistant Thermal Spray Coatings, Construction Productivity Advancement Research (CPAR) Program, US Army Corps of Engineers, Construction Engineering Research Laboratories, USACERL Technical Report 97/118, July 1997.
2. R. Sollars and A.D. Beitelman, Cavitation-Resistant Coatings for High Power Turbines, Civil Works Hydropower R&D Program, US Army Corps of Engineers, Engineering Research and Development Centre, ERDC/CERL, TR-11-21 n.d.
3. S.F. Shepelev, Status of BN-1200 Project Development, JSC Afrikantov OKBM (available at https://www.iaea.org/NuclearPower/Downloadable/Meetings/2015/2015-05-25-05-29-NPTDS/Russian_Projects/31_BN-1200_-_Obninsk_-_Shepelev_S.F._-_May_2015.pdf).
4. B.K. Sreedhar, Studies on Cavitation Erosion Resistance of Reactor Materials, PhD Thesis, Homi Bhabha National Institute (HBNI), Mumbai, India, 2016.
5. D.J. Clark and M.J. Jansen, An Overview of Magnetic Bearing Technology for Gas Turbine Engines, NASA/TM-2004-213177, ARL-TR-3254, http://gltrs.rc.nasa.gov. n.d.
6. I.J. Karassik, J.P. Messina, P. Cooper and C.C. Heald, *Pump Handbook*, 3rd Edition, McGraw Hill.
7. Y. Guojun, X. Yang, S. Zhengang and G. Huidong, Characteristic Analysis of Rotor Dynamics and Experiments of Active Magnetic Bearing for HTR-10GT, *Nuclear Engineering and Design* 237, 1363–1371, 2007.
8. L. Shi, S. Yu, G. Yang, Z. Shi and Y. Xu, Technical Design and Principal Test of Active Magnetic Bearing for the Turbine Compressor of HTR-10GT, *Nuclear Engineering and Design* 251, 38–46, 2012.
9. R. Nordman and M. Aenis, Fault Diagnosis in a Centrifugal Pump Using Active Magnetic Bearings, *International Journal of Rotating Machines*, 10(3), 183–191, 2004.
10. G. Schweitzer, Active Magnetic Bearings–Chances and Limitations, *Proceedings of 6th International IFToMM Conf. on Rotor Dynamics*. n.d.
11. B.K. Sreedhar, R. Nirmal Kumar, P. Sharma, S. Ruhela, J. Philip, S.I. Sundarraj, N. Chakraborty, M. Mohana, V. Sharma, G. Padmakumar, B.K. Nashine and K.K. Rajan, Development of Active Magnetic Bearings and Ferrofluid Seals Toward Oil-Free Sodium Pumps, *Nuclear Engineering and Design* 265, 1166–1174, 2013.
12. Y. Shimakawa, S. Kasai, M. Konomura and M. Toda, An Innovative Concept of the Sodium Cooled Reactor to Pursue High Economic Competitiveness, *Nuclear Technology*, 140, 1–17, October 2002.
13. T. Handa, Y. Oda, Y. Ono, K. Miyagawa, I. Matsumoto, K. Shimoji, T. Inoue, H. Ishikawa and H. Hayafune, Status of Integrated IHX/Pump Development for JSFR, *Journal of Nuclear Science and Technology*, 48(4), 669–676, 2011.

Appendix 1

Data on fast reactors

A1.1 INTRODUCTION

Sodium cooled fast reactors may be classified into three categories on the basis of the design intent (i.e. for research and development, technology demonstration or power generation) as: Experimental Fast Reactors, Demonstration/Prototype Fast Reactors and Commercial Size Fast Reactors. Table A1.1 gives the various categories of reactors and their countries of origin.

Table A1.1 Types of fast reactors [1]

Ser	Reactor	Type	Country
	Experimental Reactors		
1	Rapsodie	Loop	France
2	KompakteNatriumgekuhlte Kernreaktoranlage (KNK – II)	Loop	Germany
3	Fast Breeder Test Reactor (FBTR)	Loop	India
4	ProvaElementi de Combustible (PEC)	Loop	Italy
5	JOYO	Loop	Japan
6	Dounreay Fast Reactor (DFR)	Loop	United Kingdom
7	BistrijOpytnyj Reactor (BOR-60)	Loop	Russian Federation
8	Experimental Breeder Reactor (EBR-II)	Pool	USA
9	EFAPP aka Fermi	Loop	USA
10	Fast Flux Test Facility (FFTF)	Loop	USA
11	Bystrij Reactor (BR-10)	Loop	Russian Federation
12	China Experimental Fast Reactor (CEFR)	Pool	China
	Demonstration/Prototype fast reactors		
13	Phenix	Pool	France
13	Schneller NatriumgekuhlteReaktor (SNR-300)	Loop	Germany
14	Prototype Fast Breeder Reactor (PFBR)	Pool	India

(Continued)

Table A1.1 (Continued) Types of fast reactors

Ser	Reactor	Type	Country
15	MONJU	Loop	Japan
16	Prototype Fast Reactor (PFR)	Pool	United Kingdom
17	Clinch River Breeder Reactor Plant (CRBRP)	Loop	USA
18	BystrieNeytrony (BN-350)	Loop	Kazhakstan
19	BystrieNeytrony (BN-600)	Pool	Russian Federation
20	Advanced Liquid Metal Reactor (ALMR)	Pool	USA
21	Korea Advanced Liquid Metal Reactor (KALIMER-150)	Pool	Republic of Korea
22	Svinetc-VismuthBystrie Reactor (Lead-Bismuth fast Reactor) SVBR-75/100	Pool	Russian Federation
23	Bystiry Reactor EstestvennoyBezopasnosti (Fast Reactor Natural Safety) (BREST-OD-300)	Pool	Russian Federation
Commercial size fast reactors			
24	Super-Phenix I	Pool	France
25	Super-Phenix II	Pool	France
26	Schneller NatriumgekuhlteReaktor (SNR-2)	Pool	Germany
27	Demonstration Fast Breeder Reactor (DFBR)	Loop	Japan
28	Commercial Demonstration Fast Reactor (CDFR)	Pool	United Kingdom
29	BystrieNeytrony(BN-1600)	Pool	Russian Federation
30	BystrieNeytrony(BN-800)	Pool	Russian Federation
31	European Fast Reactor (EFR)	Pool	Consortium of France, UK and Germany
32	Advanced Liquid metal Reactor (ALMR)	Pool	USA
33	Svinetc-VismuthBystrie Reactor (Lead-Bismuth fast Reactor) (SVBR-75/100)	Pool	Russian Federation
34	BystrieNeytrony (BN-1800)	Pool	Russian Federation
35	Bystiry Reactor EstestvennoyBezopasnostiBREST-1200	Pool	Russian Federation
36	JNC Sodium Cooled Fast Reactor (JSFR-1500)	Loop	Japan

Table A1.2 Pump types [1]

	Experimental fast reactors	
	Main pumps	
Plant	Electrical (E) or Mechanical (M)	Main features
Rapsodie (France)	M	Centrifugal, single stage, single suction
KNK-II (Germany)	M	Centrifugal, single stage, single suction
FBTR (India)	M	Centrifugal, single stage, single suction
PEC (Italy)	M	Free surface, centrifugal, single stage, single suction
JOYO (Japan)	M	Centrifugal, Single stage, single suction
DFR (UK)	E	–
BOR-60 (Russian Federation)	M	Centrifugal, single stage, single suction
EBR-II (USA)	M / E	Centrifugal pump in primary system and AC linear induction pump in secondary system
Fermi (USA)	M	Centrifugal, single stage, single suction
FFTF (USA)	M	centrifugal, single stage, single suction
BR-10 (Russian Federation)	E	–
CEFR (China)	M	Centrifugal, single stage, single suction
	Demonstration or prototype fast reactors	
Phenix (France)	–	Centrifugal, Single stage, single suction
SNR-300 (Germany)	–	Centrifugal, single stage, single section
PFBR (India)	M	Centrifugal, single state, free surface, top suction (primary)/bottom suction(secondary)
MONJU (Japan)	M	Centrifugal, Single stage, single suction
PFR(UK)	M	Centrifugal, single stage, double suction (primary)/single stage, single suction(secondary)
CRBRP (USA)	M	Centrifugal, Free surface, single stage, double suction (primary)/single stage, single suction (secondary)
BN-350 (Kazakhstan)	M	Centrifugal, single stage, single suction.
BN-600 (Russian Federation)	M	Centrifugal, single stage, double suction(primary)/single stage, single suction (secondary)
ALMR (USA)	E	Submersible, double stator, self-cooled

(Continued)

Table A1.2 (Continued) Pump types

	Commercial fast reactors	
	Main pumps	
Plant	Electrical (E) or Mechanical (M)	Main features
KALIMER-150 (Republic of Korea)	E	Submersible double stator, self-cooled
SVBR-75/100 (Russian Federation)	M	Centrifugal, submersible
BREST-OD-300 (Russian Federation)	M	Axial single section
SuperPhénix-1 (France)	M	Centrifugal, Single stage
SuperPhénix-2 (France)	M	Centrifugal, Single stage
SNR-2(Germany)	M	Centrifugal
DFBR (Japan)	M	Centrifugal, Single stage, single suction
CDFR (UK)	M	Centrifugal, single suction, two stages(primary)/single suction, single stage (secondary)
BN-1600(Russian Federation)	M	Centrifugal
BN-800(Russian Federation)	M	Centrifugal, single stage, double suction (primary)/single stage, single suction (secondary)
EFR	M	Centrifugal, Single stage, single suction
ALMR (USA)	E	Submersible, double stator, self-cooled
SVBR-75/100 (Russian Federation)	M	Centrifugal, submersible
BE-1800 (Russian Federation)	M	Centrifugal
BREST-1200 (Russian Federation)	M	Axial single suction
JSFR-1500 (Japan)	M	Single stage centrifugal

Table A1.3 Pump locations [1]

	Experimental fast reactors			
	Location		Number of loops/pumps (for pool type reactor)	
Plant	Primary	Secondary	Primary	Secondary
Rapsodie (France)	Cold leg	Hot leg	2	2
KNK-II (Germany)	Hot leg	Cold leg	2	2
FBTR (India)	Cold leg	Cold leg	2	2

(Continued)

Table A1.3 (Continued) Pump locations

PEC (Italy)	Cold leg	Cold leg	2	2
JOYO (Japan)	Cold leg	Cold leg	2	2
DFR (UK)	Cold leg	Cold leg	24	12
BOR-60 (Russian Federation)	Cold leg	Cold leg	2	2
EBR-II (USA)	Cold leg	Cold leg	2	1
Fermi (USA)	Cold leg	Cold leg	3	3
FFTF (USA)	Hot leg	Cold leg	3	3
BR-10 (Russian Federation)	Cold leg	Cold leg	2	2
CEFR (China)	Cold leg	Cold leg	2	2
Demonstration or prototype fast reactors				
Phenix (France)	Cold leg	Cold leg	3	3
SNR-300 (Germany)	Hot leg	Cold leg	3	3
PFBR (India)	Cold leg	Cold leg	2	2
MONJU (Japan)	Cold leg	Cold leg	3	3
PFR(UK)	Cold leg	Cold leg	3	3
CRBRP (USA)	Hot leg	Cold leg	3	3
BN-350 (Kazakhstan)	Cold leg	Cold leg	5	5
BN-600 (Russian Federation)	Cold leg	Cold leg	3	3
ALMR (USA)	Cold leg	Cold leg	1	1
KALIMER-150 (Republic of Korea)	Cold leg	Cold leg	4	2
SVBR-75/100 (Russian Federation)	Cold leg	–	2	none
BREST-OD-300 (Russian Federation)	Cold leg	–	4	none
Commercial fast reactors				
Super-Phenix 1 (France)	Cold leg	Cold leg	4	4
Super-Phenix 2 (France)	Cold leg	Cold leg	4	4
SNR-2 (Germany)	Hot leg	Cold leg	4	8
DFBR (Japan)	Cold leg	Cold leg	3	3
CDFR (UK)	Cold leg	Cold leg	4	4
BN-1600 (Russian Federation)	Cold leg	Cold leg	3	6
BN-800 (Russian Federation)	Cold leg	Cold leg	3	3
EFR	Cold leg	Cold leg	3	6
ALMR (USA)	Cold leg	Cold leg	1	1

(Continued)

Table A1.3 (Continued) Pump locations

| | Commercial fast reactors | | | |
| | Location | | Number of loops/pumps (for pool type reactor) | |
Plant	Primary	Secondary	Primary	Secondary
SVBR-75/100 (Russian Federation)	Cold leg	–	2	none
BE-1800 (Russian Federation)	Cold leg	Cold leg	3	6
BREST-1200 (Russian Federation)	Cold leg	–	4	none
JSFR-1500 (Japan)	Cold leg	Cold leg	2	2

REFERENCE

1. Fast Reactor Database 2006 update, No. IAEA-TECDOC-1531, International Atomic Energy Agency, December 2006.

Appendix 2
Procedure for sodium removal from centrifugal pumps

A2.1 INTRODUCTION

In the event of maintenance of sodium-wetted equipment, such as a sodium centrifugal pump, the equipment is drained of sodium and then cleaned thoroughly using special techniques before dismantling, inspecting, and repairing of faults (if any). Removal and cleaning of the sodium-wetted equipment are challenging due to the high chemical activity and pyrophoric nature of sodium.

A2.2 METHODS FOR SODIUM CLEANING

The following methods are used for removing sodium from components removed from sodium:

1. Alcohol cleaning: The types of alcohol used for cleaning include methyl alcohol, ethyl alcohol, and butyl alcohol. The reaction rate of alcohol with sodium decreases with increasing molecular weight. While methyl alcohol and ethyl alcohol react vigorously with sodium, the reaction rate of isobutyl alcohol with sodium is mild. Alcohol cleaning is used for removing sodium from small piping and small components such as valves, fasteners, etc. Care must be exercised while using this method because of the flammability of alcohol and the evolution of hydrogen during the reaction. A disadvantage of the method is the formation of a semi-gelatinous alcoholate product on the surface of the component that results in the incomplete reaction of the sticking sodium with alcohol.

2. Steam cleaning: In this method, the sodium-wetted component is cleaned using dry steam in an inert atmosphere of nitrogen. The component to be cleaned is heated to a temperature marginally above 100°C and washed with steam under a blanket of nitrogen. The nitrogen atmosphere prevents air entry and ensures the reaction is quiet. The reaction results in the formation of sodium hydroxide with the

evolution of hydrogen. The advantage of the process is that the reaction is rapid and economical, especially for large components.

3. Water cleaning: Small components with simple geometries can be directly cleaned using a water jet or spray.

4. Thermofluid oil cleaning: Thermofluid oil such as Hytherm 500 is used to remove sodium from small parts with intricate geometry such as bellows sealed valves. The component is immersed in a bath of thermofluid maintained at 150°C–200°C, and the bath is agitated, resulting in the melting and settling of sodium at the bottom of the bath. The parts are then removed from the oil and rinsed with DM water.

5. Water vapour and carbon dioxide (CO_2) method [1, 2]: In this method, a stream of CO_2 is bubbled through water maintained at temperature of 60°C–80°C. The moisture-laden CO_2 stream is sent to the vessel containing the sodium-wetted component in an inert atmosphere of nitrogen. The wet CO_2 reacts with sodium, and the sodium hydroxide formed immediately reacts with CO_2 to form sodium bicarbonate ($NAHCO_3$) and sodium carbonate (Na_2CO_3). The hydrogen liberated during the process is a measure of the reaction rate. The flow rate of the moist CO_2 stream is controlled so that the hydrogen concentration in the process is within the safe limit (<4%). This method is preferred for reusable components because it converts corrosive sodium hydroxide, formed initially, to sodium bicarbonate, which is soluble in water. After the reaction is completed, the component is rinsed with water and dried using hot air/nitrogen.

A2.3 DESCRIPTION OF THE CLEANING FACILITY AND THE PROCESS USING WATER VAPOUR– CO_2 METHOD [1, 2]

This method is recommended for reusable components because it converts corrosive sodium hydroxide to water-soluble sodium bicarbonate, which can be removed by rinsing with water. Residual sodium hydroxide in cleaned components can result in stress corrosion cracking, and therefore, components that are to be re-used are cleaned using this method.

The basic steps in the cleaning process are:

1. Draining sodium from the pump and transferring of the pump to the vessel in the cleaning facility.
2. Reaction of the moisture-laden stream of CO_2 with the sodium adhering to the pump in an inert atmosphere of nitrogen.
3. Rinsing of the pump with water after completion of the reaction.
4. Drying using hot nitrogen.

The facility is either of the once-through type (Figure A2.1), or re-circulation type (Figure A2.2).

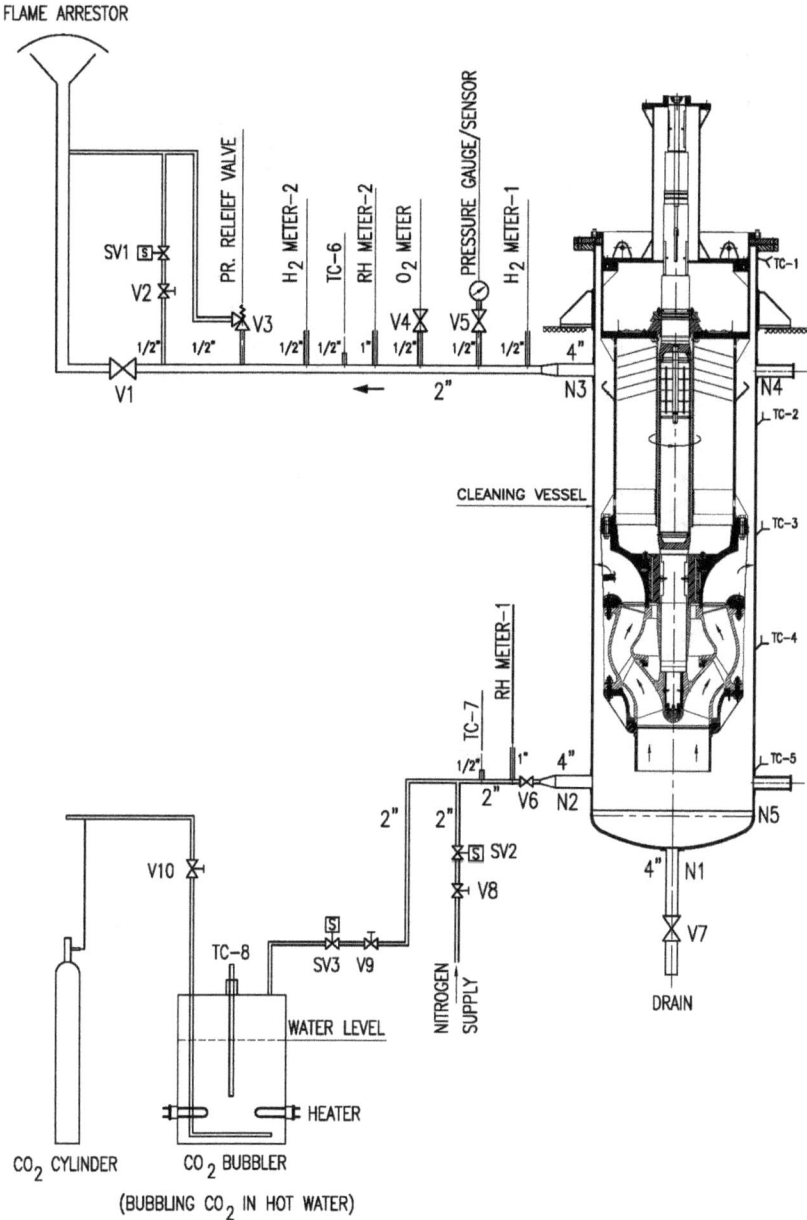

Figure A2.1 Once through type cleaning facility.

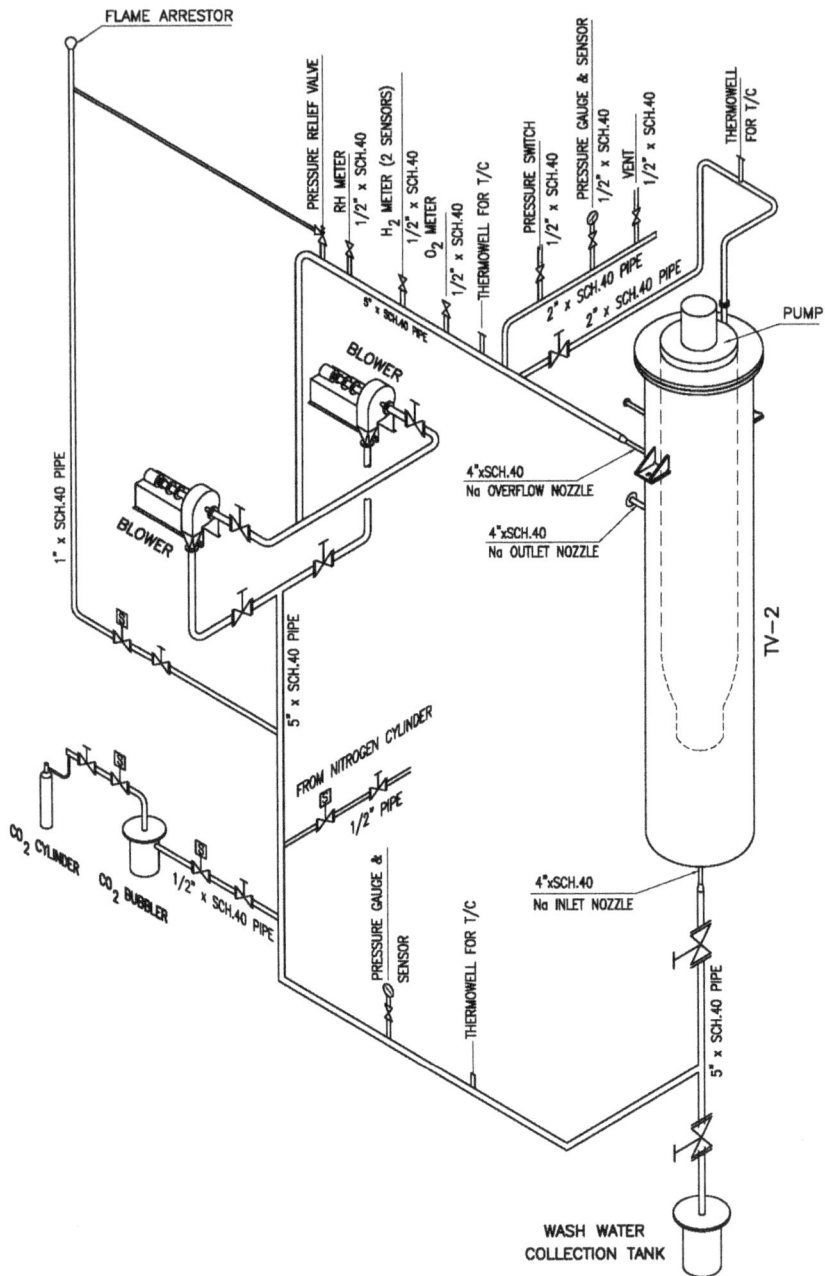

Figure A2.2 Re-circulation type cleaning facility.

The re-circulation type is preferred because it reduces the consumption of nitrogen and CO_2 gas; however, additional equipment, such as a blower with positive sealing (e.g. mechanical seal arrangement), is required.

A2.4 SAFETY PRECAUTIONS

1. All instruments and the hydrogen meters, in particular, are calibrated in the system before the start of the process.
2. Although the low explosive limit (LEL) of hydrogen in air is 4%, the maximum concentration of hydrogen in the system is limited to 3% as a precautionary measure.
3. Positive system pressure of 0.2 kg/cm^2(g)–0.3 kg/cm^2(g) is ensured during the reaction process using a feed & bleed system, and an audible alarm is provided in case the system pressure exceeds the set range.
4. Two independent hydrogen sensors operating on diverse principles are provided, and injection of moist CO_2 is stopped if either of the meter readings exceeds the set value of 3%.
5. All adhering sodium is considered to have reacted when the hydrogen meter settles at a value close to the background value.
6. In the re-circulation system, the RH meter shows a reduction in the moisture content as the reaction of moist CO_2 with the adhering sodium progresses and plateaus out when no more sodium is present, indicating confirmation of the completion of the reaction. In the once-through system, the completion of the reaction, as indicated by the hydrogen meter settling at background value, is confirmed when the RH meter on the downstream side shows no substantial decrease compared to that on the upstream side.
7. The building is well-ventilated, and the flame arrestor is located outside the building.

REFERENCES

1. S. Ignatius Sundar Raj, B. K. Sreedhar, N. Murugesan, T. G. Gunasekaran, C. Ramesh, K. K. Rajan, V. Ganesan, Sodium Removal from Sodium Wetted Under Sodium Ultrasonic Scanner, *Nuclear Engineering and Design* 254 (2013) 120–128.
2. S. Ignatius Sundar Raj, B.K. Sreedhar, B. Krishnakumar, K. K. Rajan, P. Kalyanasundaram, G. Vaidyanathan, K. Chandran, R. Sudha, N. Murugesan, C. Ramesh, P. Muralidharan, V. Ganesan, Sodium Removal and Decontamination of Fast Reactor Components, *Proceedings of international conference on peaceful uses of atomic energy*, 2009, New Delhi, India.

Index

Pages in *italics* refer to figures and pages in **bold** refer to tables.

For Product Safety Concerns and Information please contact our EU
representative GPSR@taylorandfrancis.com
Taylor & Francis Verlag GmbH, Kaufingerstraße 24, 80331 München, Germany

9 7 8 1 0 3 2 6 0 7 3 5 1